ASCE STANDARDS

American Society of Civil Engineers

Standard Guidelines for the Design of Urban Subsurface Drainage

ANSI/ASCE 12-92
ANSI Approved
March 15, 1993

Standard Guidelines for Installation of Urban Subsurface Drainage

ASCE 13-93

Standard Guidelines for Operation and Maintenance of Urban Subsurface Drainage

ASCE 14-93

Published by the
American Society of Civil Engineers
345 East 47th Street
New York, New York 10017-2398

ABSTRACT

This publication, *Standard Guidelines for the Design of Urban Subsurface Drainage, Standard Guidelines for Installation of Urban Subsurface Drainage, Standard Guidelines for Operation and Maintenance of Urban Subsurface Drainage,* contains three standards that are independent documents intended to complement the ASCE *Urban Subsurface Drainage Manual.* The first standard (ASCE 12-92) provides guidelines for design of urban subsurface drainage and covers such topics as site analysis, system configuration, filters and envelopes, hydraulics and hydrology, structural considerations, and materials. The next standard (ASCE 13-93) presents guidelines for installation of urban subsurface drainage and covers such subjects as site inspection, soil erosion, excavation, foundation preparation, and inspection of materials, equipment, and construction. The final standard (ASCE 14-93) provides guidelines for operation and maintenance of urban subsurface drainage and covers such topics as design criteria, maintenance procedures, safety, water quality, inspection, and rehabilitation.

Library of Congress Cataloging-in-Publication Data

American Society of Civil Engineers.
    Guidelines for the design of urban subsurface drainage: ANSI/ASCE 12-92, ANSI approved March 15, 1993; Standard guidelines for installation of urban subsurface drainage: ASCE 13-93; Standard guidelines for operation and maintenance of urban subsurface drainage: ASCE 14-93/American Society of Civil Engineers
        p.cm.
    Includes indexes.
    ISBN 0-87262-996-1
    1. Sewerage—Standards—United States. I. American National Standards Institute. II. Title. III. Title: Standard guidelines for installation of urban subsurface drainage. IV. Title: Standard guidelines for operation and maintenance of urban subsurface drainage.
    TD653.A62 1994   94-3853
    628'.21'021873—dc20   CIP

STANDARDS

In April 1980, the Board of Direction approved ASCE Rules for Standards Committees to govern the writing and maintenance of standards developed by the Society. All such standards are developed by a consensus standards process managed by the Management Group F (MFG), Codes and Standards. The consensus process includes balloting by the balanced standards committee made up of Society members and nonmembers, balloting by the membership of ASCE as a whole and balloting by the public. All standards are updated or reaffirmed by the same process at intervals not exceeding five years.

The following standards have been issued:

ANSI/ASCE 1-88 N-725 *Guideline for Design and Analysis of Nuclear Safety Related Earth Structures*

ANSI/ASCE 2-91 Measurement of Oxygen Transfer in Clean Water

ANSI/ASCE 3-91 *Structural Design of Composite Slabs and ANSI/ASCE 9-91 Standard Practice for the Construction and Inspection of Composite Slabs*

ASCE 4-86 Seismic Analysis of Safety-Related Nuclear Structures

Building Code Requirements for Masonry Structures (ACI 530-92/ASCE 5-92/TMS - 402-92) and Specifications for Masonry Structures (ACI 530.1-92/ASCE 6-92/TMS - 602) Specifications for Masonry Structures (ACI 530.1-92/ASCE 6-92/TMS 602)

ANSI/ASCE 7-93 *Minimum Design Loads for Buildings and Other Structures*

ANSI/ASCE 8-90 *Specification for the Design of Cold-Formed Stainless Steel Structured Members*

ANSI/ASCE 9-91 *listed with ASCE 3-91*

ANSI/ASCE 10-90 *Design of Latticed Steel Transmission Structures*

ANSI/ASCE 11-90 *Guideline for Structural Condition Assessment of Existing Buildings*

ANSI/ASCE 12-92 *Standard Guidelines for the Design of Urban Subsurface Drainage*

ASCE 13-93 *Standard Guidelines for Installation of Urban Surface Drainage*

ASCE 14-93 *Standard Guidelines for Operation and Maintenance of Urban Surface Drainage*

ASCE 15-93 *Standard Practice for Direct Design of Buried Precast Concrete Pipe Using Standard Installations (SIDD)*

FOREWORD

The Standard Guidelines for Design of Urban Subsurface Drainage is an independent document intended to complement the ASCE Urban Subsurface Drainage Manual. These publications are refined from work originally done by the ASCE Technical Committee on Drainage, which is within the structure of the Irrigation and Drainage Division. The manual will be published by the Irrigation and Drainage Division with the title ASCE Urban Subsurface Drainage Manual. There are also Standard Guidelines for Installation and Operation and Maintenance of Urban Subsurface Drainage. These standard guidelines are also developed by the Standards Committee on Urban Subsurface Drainage, which is responsible to Management Group F on Codes and Standards.

The material presented in this publication has been prepared in accordance with recognized engineering principles. These standard guidelines should only be used under the direction of an engineer competent in the field of subsurface drainage. The publication of the material contained herein is not intended as a representation or warranty on the part of the American Society of Civil Engineers or of any person named herein that this information is suitable for any general or particular use, or promises freedom from infringement of any patent or patents. Anyone making use of this information assumes all liability from such use.

ACKNOWLEDGEMENTS

The American Society of Civil Engineers (ASCE) acknowledges the work of the Urban Subsurface Drainage Standards Committee of Management Group F, Codes and Standards. This group comprises individuals from many backgrounds including: consulting engineering, research, the construction indstry, education, government, design, and private practice.

This standard was prepared through the consensus standards process by balloting in compliance with procedures of ASCE'S Management Group F, Codes and Standards. Those individuals who serve on the Standards Committee are:

Richard D. Wenberg, Chairman
James R. Andre
Virgil L. Backlund
Richard P. Baldewin
Walter R. Barker
Dennis Bauer
Robert H. Baumgardner
Mike Bealey
David E. Beck
Bishop C. Beville
Christopher B. Burke
Theodore J. Campbell
Victoria L. Daley
Barry J. Dempsey
Jan N. Dwiggins
Jeffery I. Enyart
Franchot L. Fenske
Robert S. Giurato, Secretary
James B. Goddard

Mark E. Hawley
Conrad G. Keyes, Jr.
Susan King
John M. Kurdziel
Robert D. Latta
John Meyer
Philip M. Meyers
Judith Nitsch
Walter J. Ochs
James Rabine
Alvin M. Sacks
Glen D. Sanders
Robert G. Scott
Alan N. Sirkin
K.T. Snyder, II
Ivan A. Walter
David L. Westerling
Lyman Willardson

Standard Guidelines for the Design of Urban Subsurface Drainage

Contents

LIST OF FIGURES

LIST OF TABLES

xi

Standard Guidelines for the Design of Urban Subsurface Drainage

1. Scope

The intent of this standard is to present state-of-the-art design guidance for urban subsurface drainage in a logical order. The collection and conveyance of subsurface drainage waters is within the purview of this standard for applications such as airports, roads, and other transportation systems as well as industrial, residential, and recreational areas. Incidental surface water is considered.

This standard does not address agricultural drainage, landfills, recharge systems, conventional storm sewer design, or the use of injection pumps or tubewells.

Customary units and standard international units (SI) are used throughout this document.

2. Definitions

2.1 General
This section defines specific terms for use in this standard. Reference documents 3.1.6 through 3.1.21 may be useful to augment understanding of terms in this standard.

2.2 Terms
AOS — apparent opening of geotextiles, sometimes referred to as **EOS** (effective opening size).

Aquifer — a geological formation or group of formations through which water flows.

Base drainage system — a permeable drainage blanket under a roadway, etc.

Chimney drain — subsurface interceptor drain frequently used in dams, embankments, and similar construction to control seepage within the earthen structure. Chimney drains are constructed in near-vertical orientation and discharge to outlets at lower elevations.

Colloidal fines — clay particles smaller than two microns.

Drainable water — water that readily drains from soil under the influence of gravity.

Evapotranspiration — the combined process of moisture evaporation from the soil and transpiration from plants.

Frost action — freezing and thawing of soil moisture.

Geocomposite — geosynthetic materials for collecting and transporting water while maintaining soil stability.

Geology — subsurface formations.

Geomembrane — sheet material intended to form an impervious barrier.

Geosynthetic — synthetic material or structure used as an integral part of a project, structure, or system. Within this category are subsurface drainage and water control materials such as geomembranes, geotextiles, and geocomposites.

Geotextile — a woven or nonwoven thermoplastic sheet material intended to allow the passage of water, but not fines, and without collecting fines at the soil-textile interface.

Hydraulic conductivity — see permeability.

Hydrology — movement of water in nature.

Impermeable barrier layer — a soil stratum with a permeability less than 10% of the soil permeability between the layer and the ground surface.

Longitudinal drainage system — a drainage system parallel to a roadway, etc.

Perched water table — a localized condition of free water held in a pervious stratum because of an underlying impervious stratum.

Percolation rate — downward movement of water into a soil, measured by the rate at which the water level drops in an augured hole.

Permeability — the rate at which water passes through a material.

Permittivity — a measure of the ability of a geotextile to permit water flow perpendicular to its plane.

Phreatic surface — the level of zero (atmospheric) pressure in a continuous body of groundwater.

Relief drain — any product or construction that accelerates the removal of drainable subsurface water.

Seepage — the movement of drainable water through soil and rock.

Sink — a relatively small surface depression that allows surface drainage to enter the subsurface water system.

Soil texture — the relative proportions of sand, silt, and clay particles in a soil mass.

Subsurface water — all water beneath the ground or pavement surface. Sometimes referred to as groundwater.

Transverse drainage system — a drainage system usually at some angle to a roadway, etc.

Water table — the upper limit of water in a saturated soil or underlying material.

3. References

3.1 Reference Documents

3.1.1. Sacks, A. M., "Geosynthetics," p. 14, *Remodeling Magazine*, Hanley Woods Inc., Washington, D.C., November 1987.

3.1.2. Lohman, S.W., et al., U.S. Geological Survey, "Definitions of Selected Ground Water Terms—Revisions and Conceptual Refinements," *Geological Survey Water-Supply Paper 1988*, Washington, D.C., 1972.

3.1.3. "Agri-Fabric Awareness Manual," undated, Lafayette Farm & Industry, Cuba City, Wis.

3.1.4. "Vertical Drains," undated, *GeoNotes — A ground improvements update*, Geosystems, Inc., Sterling, Va.

3.1.5. Dempsey, B.J., "Pavement Drainage System Design," prepared for Wisconsin DOT, Feb. 15–16, 1988.

3.1.6. "Concrete Pipe Design Manual," American Concrete Pipe Association, Vienna, Va., 1992.

3.1.7. "Concrete Pipe Handbook," American Concrete Pipe Association, Vienna, Va., 1988.

3.1.8. "Handbook of Steel Drainage and Highway Construction Products," American Iron and Steel Institute, New York, 1971.

3.1.9. "Clay Pipe Engineering Manual," National Clay Pipe Institute, Washington, D.C., 1982.

3.1.10. Spangler, M.G., *Soil Engineering*, International Textbook Co., Scranton, Pa., 1966.

3.1.11. "Structural Design," Section 6, *National Engineering Handbook*, Soil Conservation Service, U.S. Department of Agriculture, Washington, D.C., December 1980.

3.1.12. "The Structural Design of Underground Conduits," *Technical Release No. 5*, Soil Conservation Service, U.S. Department of Agriculture, Washington, D.C., November 1958.

3.1.13. "Standard Specifications for Highway Bridges," American Association of State Highway and Transportation Officials, Washington, D.C., 1992.

3.1.14. "Manual for Railway Engineering," American Railway Engineering Association, Washington, D.C., 1993.

3.1.15. "Airport Drainage," *ACI 50/5230-5B*, Federal Aviation Administration, U.S. Department of Transportation, Washington, D.C., 1970.

3.1.16. *Chamber's Technical Dictionary*, third edition, The Macmillan Co., New York, 1967.

3.1.17. "Nomenclature for Hydraulics," *Manual of Practice No. 43*, American Society of Civil Engineers, New York, 1962.

3.1.18. *Water and Water Use Terminology*, Thomas Printing & Publishing Co., New York, 1966.

3.1.19. *Thesaurus of Engineering and Scientific Terms*, Dec. 1967, Engineers Joint Council, New York.

3.1.20. "Drainage Manual," 1993, Bureau of Reclamation, Denver, Colo.

3.1.21. "Drainage of Agricultural Land," Section 16, *National Engineering Handbook*, 1971, Soil Conservation Service, U.S. Department of Agriculture, Washington, D.C.

3.2 General References

3.2.1. Peck, R.B., Hanson, W.E., and Thornburn, T.H., *Foundation Engineering*, 1974, John Wiley & Sons, Inc., New York.

3.2.2. Illinois Department of Transportation, "Highway Standards Manual," Springfield, Ill., November 1993.

3.2.3. Davis, C. V., and Sorensen, K. E., *Handbook of Applied Hydraulics*, 1986, McGraw-Hill, New York.

3.2.4. Meritt, F. S., *Standard Handbook for Civil Engineers*, 1983, McGraw-Hill, New York.

3.2.5. Cedergren, H. E., *Seepage, Drainage and Flow Nets*, 1967, John Wiley & Sons, Inc. New York.

3.2.6. "Hydrologic Modeling of Small Watersheds," American Society of Agricultural Engineers, St. Joseph, Mich., 1982.

3.2.7. Sowers, G. F., *Introductory Soil Mechanics and Foundations: Geotechnical Engineering*, MacMillan Publishing Co., Inc., New York, 1979.

3.2.8. Cedergren, H. R., *Drainage of Highways and Airfield Pavements*, John Wiley & Sons, Inc., New York, 1974.

3.2.9. Koerner, R. M., *Designing with Geosynthetics*, second edition, Prentice-Hall, Englewood Cliffs, N.J., 1990.

3.2.10. "Standard Test Method for Constant Head Hydraulic Transmissivity (In-Plane Flow) of Geotextiles and Geotextile Related Products," *ASTM D4716*, American Society for Testing and Materials, Philadelphia, Pa., 1992.

3.2.11. "Standard Test Method for Compressive Properties of Rigid Cellular Plastics," *ASTM D1621*, American Society for Testing and Materials, Philadelphia, Pa., 1992.

3.2.12. Beck, D. E., "Testing and Comparing Geocomposite Drainage Products," *Geotechnical Fabrics Report*, July/August 1988, Industrial Fabrics Association International, St. Paul, Minn.

3.2.13. "Specifier Manual," Advanced Drainage Systems, Inc., Columbus, Ohio, 1984.

3.2.14. "Concrete Pipe Installation Manual," American Concrete Pipe Association, Vienna, Va., 1988.

3.2.15. "Gravity Sanitary Sewer Design and Construction," American Society of Civil Engineers, *Manual No. 60*, 1982, New York.

3.2.16. Anderson, B., "Underground Waterproofing," *WEBCO*, 1983, Stillwater, Minn.

3.2.17. Driscoll, F. G., Ed., *Ground Water and Wells*, second edition, Johnson Division UOP, St. Paul, Minn., 1986.

3.2.18. Bouwer, H., *Groundwater Hydrology*, 1978, McGraw-Hill, New York.

3.2.19. Chambers, R.E., McGrath, T. J., and Heger, F. J., "Plastic Pipe for Subsurface Drainage of Transportation Facilities," Transportation Research Board, *National Cooperative Highway Research Program Report 225*, Oct. 1980, Washington, D.C.

3.2.20. Davis, S. N., and DeWiest, R.J.M., *Hydrology*, 1966, John Wiley & Sons, Inc., New York.

3.2.21. "Design of Urban Highway Drainage," *Report No. FHWA-TS-79-225*, 1983, Federal Highway Administration, Washington, D.C.

3.2.22. "Highway Subdrainage Design," *Report No. FHWA-TS-224*, 1980, Federal Highway Administration, Washington, D.C.

3.2.23. "Hydraulic Design of Highway Culverts," *Hydraulic Design Series No. 5*, 1985, Federal Highway Administration, Washington, D.C.

3.2.24. Fetter, C.W., Jr., *Applied Hydrogeology*, 1980, Charles E. Merrill, Columbus, Ohio.

3.2.25. Freeze, R.A., and Cherry, J.A., *Groundwater*, 1979, Prentice-Hall, Englewood Cliffs, N.J.

3.2.26. Powers, J. P., *Construction Dewatering*, 1979, John Wiley & Sons, Inc., New York.

3.2.27. Royster, D. L., "Landslide Remedial Measures," Tennessee Department of Transportation, Nashville, Tenn., 1982.

3.2.28. Sacks, A., "R_x for Basement Water Problems," *The Family Handyman*, Sept. 1981, St. Paul, Minn., pp. 36–40.

3.2.29. Hancor, Inc., "Recommended Installation Practice for Hancor Hi-Q Titelines, Heavy Duty and Heavy Duty-AASHTO Pipe," Findlay, Ohio, 1993.

3.2.30. Hem, J. D., "Study and Interpretation of the Chemical Characteristics of Natural Water," *1970, U.S. Geological Survey Water-Supply Paper 1473*, Washington, D.C.

3.2.31. Schuster, R. L., and Krizek, R. J., Eds., "Landslides Analysis and Control," Transportation Research Board, 1978, *Special Report 176*, Washington, D.C.

3.2.32. Schwab, G.O., Frevert, R. K., et al., *Soil and Water Conservation Engineering*," third edition,

1981, John Wiley & Sons, Inc., New York.

3.2.33. Todd, D. K., *Ground Water Hydrology*, 1980, John Wiley & Sons, Inc., New York.

3.2.34. "Handbook of PVC Pipe: Design and Construction," Uni-Bell PVC Pipe Association, 1986, Dallas, Tex.

3.2.35. "Drainage," Chapter 14, *Engineering Field Manual*, 1969, U.S. Department of Agriculture, Soil Conservation Service.

3.2.36. *Technical Guide, Section IV, Standard 606*, "Subsurface Drainage," U.S. Department of Agriculture. Soil Conservation Service, May 1988.

3.2.37. "Ground Water Manual," U.S. Department of Interior Bulletin, 1977, Washington, D.C.

3.2.38. "Evaluation of Test Methods and Use Criteria for Geotechnical Fabrics in Highway Applications," *FHWA/RD-80-021*, 1980, Federal Highway Administration, Washington, D.C.

3.2.39. Hannon, J.D., and California DOT, "Underground Disposal of Storm Water Runoff, Design Guidelines Manual," *FHWA-TS-80-218*, 1980, Federal Highway Administration, Washington, D.C.

3.2.40. "Design of Highway Drainage — The State of the Art," *FHWA-TS-79-225*, 1979, Federal Highway Administration, Washington, D.C.

3.2.41. "In-Plane Composite Drains," *Civil Engineering*, August 1984, ASCE, New York, pp. 48--51.

4. Site Analysis

4.1 General

Site analysis involves a thorough review of existing information on the site and its surrounding area. Additional studies normally are needed to fill voids in data required to prepare proper design and contract documents.

4.2 Basic Requirements

4.2.1 General. The basic information necessary for the design of all subsurface drainage facilities requires investigations of the following areas: topography, geography, water table, geology, water source, soil characteristics, environmental factors, and physical constraints.

4.2.2 Topography. All features that could influence subsurface drain location, installation, or operation must be considered in the design phase. Topographic studies of the drainage area ascertain the runoff direction following rainfalls, with lower elevations having greater drainage needs. A general topographic map of the project site and its surrounding area is required at the preliminary design stage. Topographical features of the area are essential in determining water movement to and from the drainage area. Surface water concentrations frequently result from watershed regions contributing to the area of concern.

A detailed topographic survey locating both planimetric features such as trees, ponds, ditches, culverts and catch basins, buildings, roads, walks, overhead utilities, and surface components of underground utilities, and establishing elevations of such, is necessary to develop and complete the design. The topographic information is used to establish proposed grades for the drain lines and outlet location, as well as to determine the necessity and locations of pump stations.

Additional research may be required of record information to avoid conflicts not apparent by physical evidence.

4.2.3 Geography. Geographical considerations may influence subsurface drainage system design. In contrast to most areas of the United States where special drainage is necessary to control temporary, frequently fluctuating high water tables; coastal regions typically present more extreme conditions. Generally, coastal areas have a seasonally permanent high water table, whereby the phreatic surface rises closely to the surface and remains for long periods of time. In such areas, providing adequate drainage is usually a primary construction consideration.

4.2.4 Water Table. The site may require an evaluation of the underlying water table early in the design stage. The designer must know the type of water table, the depth of the water table below the surface, and any seasonal fluctuations in elevation.

Information related to water table fluctuation throughout the year must be evaluated. An understanding of lag time for water table response after precipitation events and fluctuations related to well pumping in the vicinity of the proposed subsurface drainage site is necessary. Water table characteristics are essential when the surface of a water table is above or in close proximity to the anticipated subsurface drain outlet. It is important that the receiving drain has adequate capacity to accept this subsurface discharge continuously, particularly through periods of wet weather. Hydrostatic heads from confined aquifers, perched water tables, heterogeneous soil, and disturbed soil are important and should be noted in the site analysis reports.

With this information the designer can determine the impact the water table will have on the subsurface drain. The design of the drain must overcome any effect the water table has on the operation or efficiency of the subsurface drain.

4.2.5 Geology. Geology plays an important role in slope stability analysis where subsurface drains often are needed to stabilize slopes. Relative permeability and direction of flow in earth materials that may affect the behavior of subsurface drainage are of critical importance in subsurface drain design.

Geologic features such as rock layers, tight soil barriers, seepage paths, trench wall stability, potentials for subsidence, and collapse must be investigated.

4.2.6 Water Source. Present and projected water sources into the subsurface drain include precipitation; irrigation water and/or landscape water application; possible flood waters; canal, reservoir, or pond seepage; surface storm water; building roof runoff; and seasonally high water tables can impact the design, efficiency, and operation of the subsurface drain and must be considered.

Surface water should not be introduced into a subsurface drainage system, if at all possible, to prevent the accumulation of debris or other deleterious matter that may cause plugging of the drains and increase maintenance costs. The most cost-efficient system in terms of life-cycle costs may include separate systems to collect and drain surface water and subsurface water.

4.2.7 Soil Information.

4.2.7.1 General. Many soil profiles are non-uniform and contain layers of various soil types or rock with varying properties. Within these strata, the movement of groundwater may vary greatly. Relatively pervious soil and rock formations (aquifers) may exist to influence water movement. An aquiclude (relatively impervious stratum) also affects movement of groundwater.

Permeable soil profiles may not drain freely due to impervious soils or structures near the outer regions, thus the soil profile at the boundary of areas to be drained must be reviewed. In the soil profile study, this location of an impermeable barrier layer must be established. Subsurface drains must be placed above this layer. The efficiency of the drainage system relates to the distance between the drain invert and impermeable layer. The barrier layer may be nothing more than a change in soil texture within the profile. In practice, the barrier layer is a stratum with a permeability less than 10% of the soil permeability between this layer and the ground surface.

Specific details on soil classification, strata and layers, permeability and drainability, soil/water chemistry, temperature, and area vegetation are important design considerations.

4.2.7.2 Soil Classification. Information on soil classification can be presented in accordance with the Unified Soil Classification System, USDA Textural Classification System, or AASHTO Classification System. The primary concerns are the texture and depth of soil layers, barriers to water movement, apparent permeability and surface infiltration potential, and surface or shallow bedrock. Soil shrink-swell, cracking, subsidence, collapse, consolidation, surface sealing, and compaction potentials are also important factors. The USDA Textual Classification System has some preferred details for use in Section 5 but all systems named previously will help in determining classifications.

4.2.7.3 Strata and Layers. The soil strata and layering require identification since they affect the water movement to and the installation of the subsurface drain lines. Depth, thickness, permeability, slope, and extent of the layers are required data.

4.2.7.4 Permeability and Drainability. These factors should be determined through hydraulic conductivity tests. Performance of in situ saturated hydraulic conductivity tests are recommended.

4.2.7.5 Soil/Water Chemistry. Salinity, corrosivity, and pH must be determined before selecting pipe and pump materials. Any substances present or projected at the site that could impact either material selection or disposal of the drain effluent must be identified. Substances such as iron pyrites, sodium, calcium, selenium, boron, arsenic, or iron may create disposal and/or material selection problems.

4.2.7.6 Temperature. An understanding of the temperature range and fluctuations is important for evaluating potential frost action or freeze-thaw. Frost action occurs when moisture is in the freezing zone of the soil and is caused by the freezing and thawing of soil moisture. Frost action may influence material selection and installation depth.

4.2.7.7 Vegetation. Trees and shrubs are particularly important design factors. Water-seeking vegetation, such as the willow tree, is critical in subsurface drain location since the roots tend to plug drains. Other vegetation and landscape details must be noted to provide information on rooting depths that are important to drain depth and spacing.

5

4.2.8 Environmental Factors.

4.2.8.1 General. Some of the major environmental considerations in the design of subsurface drainage systems are listed later. These factors must be considered in design to prevent adverse environmental impacts to adjacent parcels of land, residents, and environmentally sensitive ecosystems:

a. water quality; b. flooding; c. wetlands; d. principal or primary aquifers; e. hydrology.

4.2.8.2 Water Quality. Design, construction, and operation of subsurface drainage systems and components must be performed in such a way as to not contaminate nearby channels. Contamination occurs when subsurface water accumulates excessive minerals, pesticides, or herbicides, as the water passes through the soil. Water tables crucial to agricultural farm applications and/or those depended upon for sources of potable water must be of high quality. Certain states regulate the possibilities of contamination through required permits for discharge. Such permit requirements must be adhered to in design, construction, and operation. Employing oil/water separation facilities in parking areas and the selective use of discharge points are some of the considerations a designer can use to mitigate possible contamination of water quality.

4.2.8.3 Flooding. In areas where projects are diverting a high flow rate of water off-site, it is imperative for the designer to consider the possibility of flooding. Flooding is caused by overflowing streams and runoff from adjacent properties. Potentials of flooding must also be examined in situations where subsurface drainage is diverted into certain channels such as ditches, enclosed drainage networks, and/or streams. Sufficient capacity for the subsurface flow should be available in the channel during storms. Often an analysis of downstream properties must be performed to determine any flood hazards. The potential for downstream flooding is not always considered in adequate detail and may lead to liability concerns after the project is constructed.

4.2.8.4 Wetlands. Additional consideration is required for the ecological-environmental aspects of the site when applying artificial drainage to a wetland area.

Wetlands are classified by local, state, and federal governments. Most wetlands are considered to be environmentally sensitive ecosystems. The maintenance of these ecosystems is important. Many wetlands filter natural and man-made pollutants. Changes in the quality and/or quantity of subsurface drainage waters entering a wetland can adversely affect this sensitive filtering process and may cause detrimental effects to flora and fauna associated with the wetland.

Any project associated with a wetland will most likely require permits from local, state, and/or the federal government (Section 10 discusses permits and codes).

4.2.8.5 Principal or Primary Aquifers. These aquifers are often tapped as a main water source. The intended use of this water determines the necessity and amount of protection required. If a possibility exists for contamination of the aquifer, mitigative measures must be taken to prevent such contamination. Other design alternatives including relocation of the system or a treatment and monitoring program of the subsurface discharge may be necessary to remove the contamination potential.

4.2.8.6 Hydrology. Hydrology describes the movement of water. Development intervenes the natural hydrological cycle and generates an artificial water cycle. Although this has typically been an insignificant design factor, it has grown to be a valid concern in recent years. The development of subsurface drainage systems should follow the natural hydrological cycle as closely as possible. For example, if the natural cycle exists as rainfall percolating into groundwater, then joining surface water courses, the man-made cycle should parallel this movement. Not all hydrological cycles are this simple and thus easily parallelled. It is important to consider the natural or existing hydrological cycle of the site in the design.

4.2.9 Physical Constraints.

Most urban settings have constraints related to existing or planned utilities that must be considered in the design of subsurface drainage systems. Compatibility of proposed systems with existing drain systems is critical to any layout. The location of utilities may require special consideration in the design stage to accommodate pump stations, future development, and master planning.

Additional physical constraints can be identified through a topographic survey, as discussed in Section 4.2.2.

5. System Configuration

5.1 General

An urban subsurface drainage system will include any or all of the following components: collection and conveyance lines, outlets, and appurtenances.

5.2 Collection System Types

5.2.1 Pipes may be used for both collection and conveyance, and in this standard are intended only to receive the liquid (usually water) to be drained. Modern pipe materials include concrete, ductile iron, steel, plastic, and clay, and the pipe may have a solid or perforated wall. Pipes are joined in various ways to provide soil-tight or watertight joints. Connector devices include tees, wyes, elbows, and adapters for different diameters and materials.

5.2.2 Geocomposites of similar or different materials may be used as collection devices to create an "in-plane" envelope for intercepting liquid flowing at right angles to the envelope. Geocomposites may also be used to convey liquid to an outlet. Geocomposites may be installed vertically, i.e., attached to the exterior of a structural foundation wall or horizontally in a trench.

5.2.3 Geomembranes may be used as a barrier to liquid flow to waterproof foundation walls, seal under pavements, cut off fills, and line cuts, e.g., in earth dams, hazardous waste dumps, drainage ditches, and retaining walls. Geomembranes are manufactured in a variety of materials, including plastic, synthetic rubber, and asphaltic compounds.

5.2.4 Geotextiles may be used as coverings and liners in several construction applications. In drainage applications, geotextiles serve as filters that pass water and colloidal fines while restricting soil migration.

5.2.5 Aggregates may be used as filters or envelopes. Aggregates may be either sands and gravels or crushed stone. Aggregates should be essentially free of sediment and foreign materials.

5.2.6 Wick Drains may be used as vertical drainage to abet the upward flow of water from underground sources and make it free flowing on the surface. Wick drains accelerate the consolidation of soft and compressible soils.

5.3 Conveyance and Outlet

Water collected in a drainage system is normally conveyed to a safe and adequate outlet, such as a natural outfall or storm drainage facility. Where gravity flow is not feasible, pumping is necessary. Most of the collection system types in Section 5.2 may also be used in conveyance systems, which have different design and installation requirements.

5.4 Applications

This section includes, but is not limited to, the major applications of the collection system types described in Section 5.2. Combinations of these systems may be used in a drainage system. Geotextiles typically encapsulate aggregate to form a viable drainage system.

5.4.1 Foundation Drains have application to buildings, bridges, dams, and retaining walls, where structural elements are involved and removal of water is needed.

All collection system types have application in foundation drains. In many foundations, a subsurface drainage system using pipes, geocomposites, geomembranes, geotextiles, and aggregates is used. Geomembranes by themselves will keep a structure dry but do not remove the subsurface water. The use of wick drains is limited to areas where the hydraulic head (soil pore pressure) is great enough to force subsurface water to the surface. A wick drain system can have maintenance problems where surface grading is slight and freeze/thaw cycles are a reality. See Figure 5.1 for typical applications of foundation drains.

5.4.2 Roads, Railroads, and Airports have pavement facilities defined as roadways, railroads, taxiways, or runways. The typical pavement system will have significant length versus width ratio. The types of drainage systems typically are longitudinal, transverse, and base.

5.4.2.1 Longitudinal Drainage System. A longitudinal system is essentially parallel to the roadway or railroad. It is usually placed at a depth that allows the road section to be gravity drained. Pipe systems, geotextiles, aggregates, and geocomposites are commonly used for longitudinal systems. When curb and gutter are present for a roadway system, discharge into a surface drainage system is usually required. See Figure 5.2 for typical applications of the longitudinal drainage system.

5.4.2.2 Transverse Drainage System. A transverse system is one that drains across the roadway or railroad. It is usually at right angles to the roadway but can be placed at any angle. A transverse drainage system consists of pipe, geotextile, aggregate, and/or geocomposite. The main need for this system is in sections where problems with groundwater are anticipated. This system should connect to the longitudinal subsurface system or the surface drainage

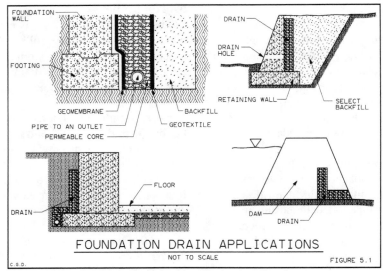

Fig. 5.1 Foundation Drain Applications

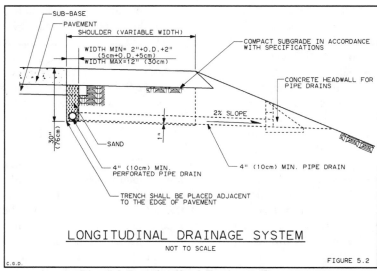

Fig. 5.2 Longitudinal Drainage System

system, such as the ditch. Figures 5.3 and 5.4 show typical transverse drainage systems.

5.4.2.3 Base Drainage System. A base system is one that usually consists of a permeable drainage blanket under the entire roadway or railway. This system can also add to the structural integrity of the pavement. A properly graded aggregate is often sufficient for a base drainage system. Geotextiles are often used with an aggregate to provide separation and filtration or strength. A properly crowned roadway, railroad, taxiway, and runway subbase will en-

hance the drainage system. Figure 5.5 shows a typical base drainage system.

5.4.3 Parking and Other Paved Areas, such as parking lots, turf fields, and playgrounds, have pavement systems where the length versus width ratio approaches one (1). The type of drainage needed can vary with the paved surface. Asphalt, stone, and artificial turf will be considered as pervious surfaces. Concrete is considered to be a relatively impervious surface.

8

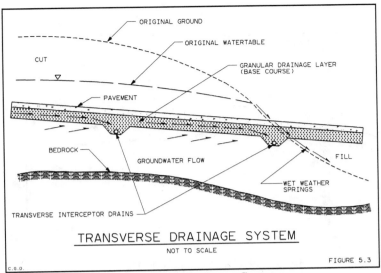

Fig. 5.3 Base Drainage System

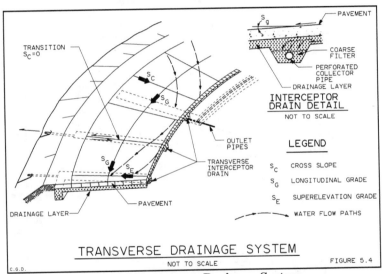

Fig. 5.4 Transverse Drainage System

5.4.3.1 Pervious Facilities. These facilities can be large with numerous low points where surface and subsurface drainage collect. Accumulated drainage flows through the pervious surface to the subsurface collector system of pipes, geocomposites, and aggregates.

The more pervious the facility, the greater the flow to the subsurface system. In general, infiltration rates are high. Often these systems are constructed in tandem with a surface drainage system.

Geotextiles and geomembranes can be used depending on effects of the water table. The most important design consideration is to properly design the subbase to get subsurface drainage into the pipe system. Because of the various rates of flow, a thorough study is needed before the subsurface system is designed for pervious facilities. Figure 5.6 shows a typical pervious system.

5.4.3.2 Impervious Facilities. A large impervious facility, usually concrete pavement, typically

9

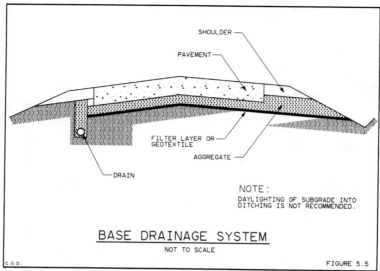

Fig. 5.5 Transverse Drainage System

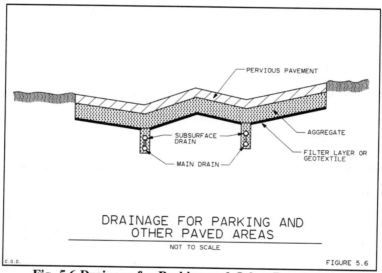

Fig. 5.6 Drainage for Parking and Other Paved Areas

has numerous low points to collect surface and subsurface drainage. However, the volumes of subsurface water are usually very small due to low rates of infiltration through the pavement. Perforated pipes, aggregates, geocomposites, and geotextiles are used in these types of facilities. Unless a rising water table is present, geomembranes are seldom used because there is rarely a need to keep an area completely dry. Figure 5.7 shows a typical impervious system.

5.4.4 Recreation and Turf Areas, such as parks, golf courses, and athletic fields, are also used for detention or retention of surface water. Because of the relatively large size of these facilities, the cost of a subsurface drainage system can be an issue. The system used should be designed for low maintenance

10

and take into account variations of the surface over time. All the collection systems are applicable in one form or another, although the importance of keeping these areas "dry" should be weighed before a system is designed. For example, a revenue-producing golf course needs to drain well to be back in service as soon as practicable after a storm ends. Figure 5.8 shows a typical application for a recreation and turf area.

5.4.5 Landscape Areas are usually situated in urban areas and are typically small areas in relation to the overall development. Large landscape areas

are usually part of an overall recreational area. The importance of keeping the areas "dry" must be weighed before a system is designed. For example, an area that has a large amount of foot traffic will need to be back in service soon after the storm ends.

All collection systems, except wick drains, are applicable. Wick drains are not used because their resultant surface drainage runoff is unacceptable in an urban area. Geomembranes may be used in planter areas to keep foundations dry. Pipe systems may be used in tandem with city drainage systems. Geocomposites, geotextiles, and aggregates may be used to

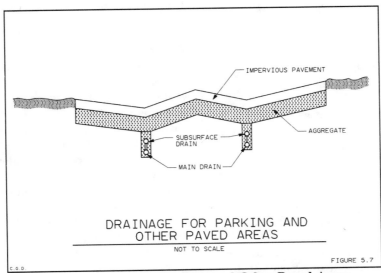

Fig 5.7 Drainage for Parking and Other Paved Areas

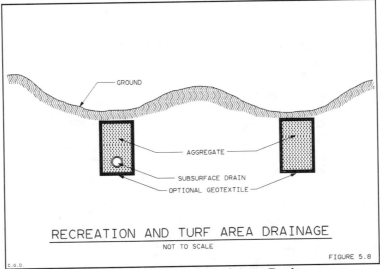

Fig. 5.8 Recreation and Turf Area Drainage

11

remove surface ponding. The amount of drainage provided should be related to the effect of ponding on the landscaped area.

5.5 Appurtenances

An urban subsurface drainage system may include various appurtenances necessary for a complete and operational system including such items as lift stations, pumping stations, vaults, manholes, and cleanouts.

5.5.1 Pumping Stations may be used in conveyance systems to transport water to a distant and higher discharge outlet. Pumping stations normally include pumps, piping, valves, ducts, vents, controls, electrical equipment, and accessories.

5.5.2 Vaults may be used in any drainage system to house electrical or other equipment underground. Vaults normally include ducts, piping, valves, vents, and accessories in addition to the equipment being housed.

5.5.3 Manholes may be used in conveyance systems to facilitate inspection and maintenance of the drainage pipe. In small-diameter pipe systems that cannot be entered by personnel, manholes are normally constructed at each change in grade, pipe size, or alignment, and at intervals for cleaning purposes. In pipe systems that can be entered by personnel, changes in alignment may be affected by curved pipe to eliminate the need for a manhole. Manholes may be constructed of concrete or other approved materials. Precast concrete or prefabricated manhole units are joined in various ways to provide soil-tight joints. Pipe-to-manhole connections should be soil-tight as required and provide flexibility at the pipe-manhole interface. A watertight connection may be required in some installations. Manholes are normally capped with a metal casting with a removable lid or with a concrete slab that includes the metal casting and removable lid.

5.5.4 Cleanouts may be used in conveyance systems to facilitate inspection and maintenance of drainage systems that cannot be entered by personnel and are not scheduled for frequent inspection and maintenance. Cleanouts are normally constructed at grade and alignment changes of approximately 45 degrees or greater. Cleanouts are usually a wye section in the pipeline, with a removable stopper in the wye. Cleanouts in public rights-of-way are normally extended to a point 6 to 12 inches below finished ground surface and plugged with a removable stopper. Cleanout wyes should be the same material as the main pipeline. Cleanout extensions may be of any approved pipe material.

6. Filters and Envelopes

6.1 General

6.1.1 Filters are permeable materials placed around the drainage pipe or other drainage product to restrict migration of noncolloidal particles from the surrounding soils.

6.1.2 Envelopes for drains are permeable materials placed around the pipe or drainage product to improve flow conditions in the area immediately around the drain and for improving bedding and structural backfill conditions.

6.2 Determination of Need

6.2.1 Filters are required in soils with piping or migration potential. Generally, Table 6-1, from the Soil Conservation Service, provides good guidance based on soil classification. Other considerations, such as perforation size in the pipe and velocity of flow must be considered. The filtration requirements take precedence over considerations for the need for an envelope.

6.2.2 Envelope characteristics should be considered in most applications. This is true for all pipe types for structural purposes and to increase water flow rates and, possibly, water storage capacity. Refer to Table 6-1 for guidance.

6.3 Design of Filters and Envelopes

6.3.1 Filters. In selecting a proper filter medium, it is necessary to determine the gradation curve, using wet sieve analysis techniques of the in situ material to be drained. This should be done from actual site samples, because gradations for a given soil type can vary widely. From these curves the following design calculations establish limits that should be met by the filter material.

$$\frac{D_{15} \text{ (filter)}}{D_{85} \text{ (soil)}} < 4 \text{ to } 5 < \frac{D_{15} \text{ (filter)}}{D_{15} \text{ (soil)}}$$

where, from the gradations: D_{15} = 15% size (15% is finer) and D_{85} = 85% size (85% is finer).

The left half of the equation is the piping protection ratio. The right half of the equation is for the permeability ratio in the system.

Not more than 5% of the filter material should pass the #200 sieve or D_5 (filter) > 0.075 mm. This is for

Table 6-1. A Classification to Determine the Need for Drain Filters or Envelopes, and Minimum Velocities in Drains

Unified Soil Classification	Soil Description	Filter Recommendation	Envelope Recommendation	Recommendations for Minimum Drain Velocity
SP (fine) SM (fine) ML MH	Poorly graded sands, gravelly sands. Silty sands, poorly graded sand-silt mixture. Inorganic silts and very fine sands, rock flour, silty or clayey fine sands with slight plasticity. Inorganic silts, micaceous or diatomaceous fine sandy or silty soils, elastic silts.	Filter needed	Not needed where sand and gravel filter is used but may be needed with flexible drain tubing and other type filters.	None
GP SC GM SM (coarse)	Poorly graded gravels, gravel-sand mixtures, little or no fines. Clayey sands, poorly graded sand-clay mixtures. Silty gravels, poorly graded gravel-sand silt mixtures. Silty sands, poorly graded sand-silt mixtures.	Subject to local on-site determination.	Not needed where sand and gravel filter is used but may be needed with flexible drain tubing and other type filters.	With filter—none. Without filter—1.40 feet/second. (0.43 m/s)
GC CL SP, GP (coarse) GW SW CH OL OH Pt	Clayey gravels, poorly graded gravel-sand-clay mixtures. Inorganic clays of low to medium plasticity, gravelly clays, sandy clays, silty clays, lean clays. Same as SP & GP. Well-graded gravels, gravel-sand mixtures, little or no fines. Well-graded sands, gravelly sands, little or no fines. Inorganic, fat clays. Organic silts and organic silt-clays or low plasticity. Organic clays of medium to high plasticity. Peat.	None	Optional. May be needed with flexible drain tubing.	None—for soils with little or no fines. 1.40 feet/second for soils with appreciable fines. (0.43 m/s)

U. S. DEPARTMENT OF AGRICULTURE
SOIL CONSERVATION SERVICE

Source: Ref 3.1.2, "Drainage of Agricultural Land."

protection of the permeability of the filter.

6.3.1.1 Pipe Holes, Slots, and Joints. Filter materials (or in situ soils if filters are omitted) must be coarse enough not to enter the pipe openings. The following equation should be used:

$$\frac{D_{85} \text{ (filter)}}{\text{max. opening in pipe}} \geq 1$$

6.3.1.2 In cases where these ratios cause some difficulty in selecting a filter satisfying both the in situ soil/filter relationship and the filter/pipe relationship, multiple filter layers may be required. In such a case, the recommended ratios in 6.3.1 must be met at each interface.

6.3.2 Envelopes may be used to provide proper bedding or for improved hydraulics flow into or around the drainage pipe. The same design criteria used for natural filters (6.3.1) is customary but may be more expensive than necessary if a well-graded coarse sand–gravel material will meet the need.

6.4 Filter Materials

6.4.1 Natural Materials.

6.4.1.1 Aggregates for drainage filters must be selected and sized for maximum permeability and the filter criteria given in Paragraph 6.3.1. Aggregates must be chemically and structurally stable. Flow rates through aggregates vary greatly depending upon gradation.

6.4.1.2 Graded sands may be used as filters if their permeability is sufficient and they meet the filter criteria in Paragraph 6.3.1.

6.4.1.3 Blast furnace and steel mill slags, although not truly natural, belong in this group. Again, sizing and permeability are critical. These materials must be regarded with caution due to the chemical reaction potential of certain of these slags. In some cases, these materials can be cementatious. They can also release materials, which can clog drains. Structural stability may also be a concern.

6.4.2 Geotextiles and Synthetic Filters. Geotextiles used in filter applications should be nonwoven needle-punched, knitted, or spun bonded. Critical parameters for filter applications are permittivity, A.O.S. (or E.O.S.), and survivability during construction. Geotextiles are available and manufactured from polypropylene, polyester, or nylon.

Permittivity values in excess of 1 sec^{-1} should be required. Permittivity, not permeability, should be used because it is more important to evaluate the quantity of water that would pass through the fabric under a given head over a given cross-sectional area without regard to fabric thickness.

A.O.S. of a fabric is the number of the U.S. standard sieve with openings closest in size of the geotextile openings. The soil/fabric relationship should be:

$$\frac{D_{85} \text{ (soil)}}{\text{A.O.S. of fabric}} \geq 1$$

Two properties critical to the survivability of a geotextile are grab strength and elongation. Tested per ASTM D-1682, grab strength should exceed 100 pounds and grab elongation should exceed 50%

7. Hydraulics and Hydrology

7.1 General Information

Successful design of subsurface drainage systems requires an understanding of the behavior of groundwater (hydrology) and the effects of drainage systems on saturated soils. Furthermore, understanding groundwater collection and discharge properties (hydraulics) of drainage systems is essential to ensure design and performance.

7.2 Water Sources

7.2.1 Subsurface Water Sources. In this document, subsurface water is considered to be all water beneath the ground or pavement surface and will be sometimes referred to as groundwater.

Soil water is generally of three types: drainable water, plant-available water, and unavailable water. Plant-available water is often referred to as "capillary water," since it is retained by the soil in small soil pores where capillary forces prevent gravity-influenced drainage and is available for plant root absorption.

Drainable water may be considered to be water that readily drains from soil under the influence of gravity. Drainable water moves through soils in direct proportion to the soil's permeability and hydraulic gradient, thus low permeabilities result in slow natural drainage of saturated soils.

Unavailable water is held tightly in thin fi'•ns surrounding individual soil particles. The strong film bond makes this water nondrainable and unavailable

14

to the vegetation. The amount of this hygroscopic water varies with the surface area of the soil particles and, therefore, is highest in clay and organic soils.

Most subsurface water results from surface infiltration, although water can enter the subsoil from adjacent areas. Another potential contributor to excess soil wetness is a perched water table that generally forms above an impermeable soil layer.

Water infiltration in soils is governed by soil type, season of the year, degree of soil moisture content at time of rainfall or irrigation, type and extent of vegetative cover, surface "crusting" tendency from rainfall impact, and characteristics of the particular rainfall event.

7.2.2 Surface Water Sources. Water from a rainfall or irrigation event that does not infiltrate the soil appears as surface water. An exception to this generalization is a condition of interflow, wherein infiltrated water moves along an impermeable strata and exits the soil mass at a hillside or cut.

Surface water becomes a consideration in subsurface drainage analysis when it becomes runoff or interflow to the drainage area under study and contributes to the anticipated water removal requirements of the subsurface drainage system. Surface water runoff is a major concern in urbanized areas, where development results in a high percentage of impervious surfaces such as roofs, driveways, and streets. In evaluating the subsurface water removal requirements of a specific area, adjacent areas that represent potential watersheds must be considered. Urban watersheds usually have greatly reduced water absorption and interception capacity, resulting in significant surface water discharge quantities.

Surface water may be free to flow to adjacent areas (runoff) and contribute to soil saturation in another zone and/or streamflow. Some surface water is retained on the ground surface in depressions which, if soil permeability is extremely low, will evaporate or pond.

7.3 Establishing the Need for Subsurface Drainage

7.3.1 General. Excess soil moisture or ponded surface water may be caused by one or more factors. Disregarding runoff from adjacent areas or subsurface aquifers, low soil permeability is the usual cause of extended water retention following precipitation.

Permeability, or hydraulic conductivity, is used as a measure of the soil's ability to transmit water by gravity. Generally, coarse materials such as sand are highly permeable and have good transmission rates. Clay soils, however, are usually relatively impermeable and water retention is long-term in the absence of a drainage system. Crude estimates can be made of a soil's permeability by realizing that the passage of water depends greatly upon voids in the soil structure, explaining chiefly why granular soils with higher void sizes move water better than compact soils with small grain sizes or void sizes.

Soil permeability is best determined by careful laboratory or on-site analysis. Because of the complex nature of soil composition and its influence on permeability, it is virtually impossible to closely establish this property without direct measurement. While laboratory testing (ASTM D2434) is highly recommended, the accuracy is directly influenced by the care taken in reconstructing the natural soil state. Some on-site measurements closely determine soil permeability and those techniques that evaluate the soil to depths below the drains should be considered most reliable.

7.3.2 Removal Criteria for Different Environments and Climates. Subsurface drainage, when viewed as water management, is used agriculturally as a method to improve the soil environment for plants and to eliminate operational hazards and nuisances. The benefits are economical and safe in nature. The rate at which surplus groundwater is removed relates primarily to the moisture/air requirements of vegetation.

Drainage of urban areas generally relates to lawns, parks, and recreational turf, and usually involves faster removal rates than for agriculture. Faster water removal from the surface and top soil zone allows quicker access to turf surface and minimizes athletic disturbance of vegetation. Aeration of vegetation is essential to health and durability, and some areas require enhanced drainage for salinity control. Groundwater control is necessary for the correct performance of septic tank absorption fields.

Climatic conditions must be considered. Soils in humid regions often require more extensive drainage systems than soils in arid regions. Temperature and humidity conditions interact with soil characteristics to influence moisture control requirements.

Comprehensive data for the selection of optimum drawdown rates for nonagricultural drainage systems are not yet available. Local turf specialists should be of assistance, and regional data should identify distinctive climatic conditions such as humidity and evapotranspiration rates.

7.3.3 Special Requirements for Paved Surfaces.

Vehicular traffic on pavements with saturated subbase results in early deterioration. Water enters through openings in the pavement, from high groundwater conditions or as seepage from adjacent regions. Correctly designed paved areas should have highly permeable base or subbase construction and drainage systems for promoting rapid outflow of infiltrated water.

7.4 Basic Subsurface Drainage Theory

7.4.1 Controlled Water Table Elevation.

Subsurface drainage is accomplished by placing an artificial channel below the water table so that the hydraulic head of the channel is less than that of the soil to be drained. The hydraulic head differential creates a hydraulic gradient in the direction of the artificial channel, depressing the phreatic line (also called free water surface; see Figure 7-1) in the vicinity of the artificial channel. The constant removal of water flowing into the drainage sink maintains the hydraulic head differential, thus maintaining the depressed phreatic line.

The hydraulic gradient and the hydraulic conductivity of the soils to be drained govern the rate at which water moves toward the sink. Control of water is accomplished by controlling the hydraulic gradient. Therefore, flow is regulated by adjusting the depth of the sink and the spacing between sinks, and by locating the sinks to take advantage of the more permeable soils in the area to be drained.

7.4.2 Drainable Versus Nondrainable Groundwater.

Drainable water, or "gravity water," is that which is free to move through the soil by the force of gravity or hydraulic gradients. When all pores are filled with water, the soil is considered to be "saturated."

Some water is held in the soil against gravity. It includes the film of water left around the soil grains and the water filling the smaller pores after gravity water has drained off. This water is considered to be nondrainable groundwater.

7.4.3 Drainage Formulas and Recommended Practices.

Selection of the proper drainage system to control the anticipated flow will vary with the specific application, but in all applications the procedure begins with a determination of the allowable subsurface water elevation and the drain depth. Spacing of drains can then be calculated and a discharge computed for each drain.

Most methods for estimating drain spacing are empirical and were developed to meet specific char-

acteristics of a particular area. Some methods are based on assumptions of steady-state flow conditions where the hydraulic head does not vary with time. Other methods assume transient flow conditions where the hydraulic head changes with time.

The most widely used steady-state formula for subsurface drainage spacing is the ellipse equation or the modified ellipse equation where convergence of flow at the drain is considered significant. In using the ellipse equation, the depth to the drain is first established, then spacing of the drains is computed by the formula. The ellipse equation is based on an assumption that the streamlines of flow in a gravity drainage system are horizontal and the velocity of flow is proportional to the hydraulic gradient or the free water surface. Although approximate, these assumptions may closely approach actual conditions in certain sites. For this reason, use of the ellipse formula should be limited to the following conditions:

(1)   Where the groundwater flow is known to be largely in a horizontal direction, such as stratified soils with relatively permeable layers acting as horizontal aquifers.

(2)   Where soil and subsoil materials are underlain by an impermeable barrier at relatively shallow depths (less than twice the depth to the drain), which restricts vertical flow and forces the groundwater to flow horizontally toward the drain.

(3)   Where open ditches are used or where drains with sand and gravel filters or porous trench backfill are used. These conditions cause a minimum of restriction to flow into the drains and minimize convergence of flow at the drains.

The ellipse equation is usually expressed in the form:

$$S = \sqrt{\frac{4K_s \ (m^2 + 2am)}{q}}$$

where: S = drain spacing feet (m); K_s = average hydraulic conductivity, in./hr. (m/day); m = vertical distance, after drawdown, of water table above drain at midpoint between drain lines, ft (m); a = depth of barrier layer below drain, ft (m); and q = drainage coefficient, in./hr (m/day). Details of the symbols and parameters for Equation 7.1 are shown on Figure 7.1.

Other subsurface water drainage formulas have been developed, the most notable of which are those

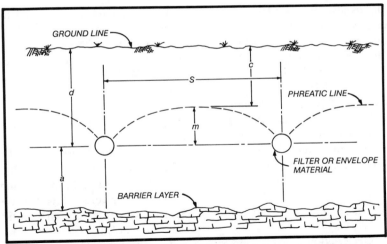

Fig. 7.1 Cross-Sectional View Showing Symbols Used in Ellipse Equation. Note: Dimensions c and d are distances from the surface to the permanent water table elevation and to the invert of the drainage system respectively. These values are controlled by the factors a, m, and s in the ellipse equation. Source: Ref. 3.1.21, "Drainage of Agricultural Land, U.S.D.A. SCS."

developed by Ernst, Hooghoudt, and Dumm. These drainage models are considerably more complicated and apply to more complex types of geological and hydrological conditions. The particular formula selected for computing the spacing of relief drains is influenced by site conditions and experience from drains in similar soils and climates.

One method of accounting for the losses due to convergence was developed by Hooghoudt. This method uses an equivalent depth to the barrier below the drain and uses this depth in the ellipse equation. The procedure for computing the adjustment for convergence is discussed in U.S. Department of Interior and U.S. Department of Agriculture references (see Chapter 3) with examples presented for use of both the ellipse and modified ellipse equations.

7.5 Subsurface Drain Applications

7.5.1 Soil Consolidation Using Wick Drains.

Compressible soils with high water content are generally unsuitable on construction sites selected for any structure requiring a stable foundation. Since the damaging effects of soil settlement are well documented, the practice of soil stabilization through the process of consolidation has been widely accepted. This process fundamentally involves the development of a surcharge loading on the site to produce excess pore pressures in the water, forcing water from the

soil pores. According to the theory of consolidation, the results vary inversely with the soil permeability and directly to the square of the longest drainage path. In most cases, especially where fine silts and clays are present, the process is extremely long if unaided by a drainage system.

Establishing vertical holes and filling them with sand or inserting geocomposite wicks (flexible plastic core with geotextile overwrap) deeply into the soil to be consolidated shortens the drainage path and greatly accelerates the rate of consolidation. This permits quicker and more reliable construction.

The design of sand drain systems—more recently being replaced by smaller geocomposite wicks—has evolved as a precise science since its beginning in 1926.

7.5.2 Groundwater Removal with Relief Drains.

Relief drains are broadly considered to be any product or construction that accelerates the removal of drainable subsurface water. Such drains are commonly required in soils whose properties result in extended saturation periods following rainfall events, thus affecting infiltration and runoff, vegetative health, and seepage forces on underground structures. In the broadest sense, relief drains can take the form of trenches or paved channels, but are more commonly a buried product or structure that is systematically designed and located to serve as a subsurface water collector or serve as both a surface and subsurface

17

water collector. Relief drains also discharge the collected water to a selected area or structure.

Relief drain design and construction has evolved over many centuries to include sand- and aggregate-filled trenches (both with and without perforated pipe in the lower portion of the select fill), pipe with small perforations through the wall surfaces (installed either with or without a geotextile covering), solid-wall pipe installed with gaps between each length (called "open joint" pipe), and geocomposite fin drains (formed plastic interior members overwrapped with filter textiles). Various configurations of these products and systems can provide groundwater removal from turf areas and protect subsurface structures from seepage forces (see Figure 7-2).

Drainage systems must serve two primary functions over long periods of time: collect groundwater and then discharge the collected water to a selected point. In nearly all cases, the motivation is gravitational force and the resulting driving energies are relatively small, requiring care in design and placement to avoid major performance reduction.

The discharge of collected water within the relief drain follows traditional hydraulic flow criteria for pipes or permeable sand/aggregate drains, generally employing techniques established by Chezy, Manning, or others (for tubular drains) and Darcy's laws (for sand or aggregate drains). In all cases, the liberal use of safety factors is encouraged because of the inherent imprecision in seepage design.

Water flowing into the collector drains may transport fine soil particles. The movement of fines is dependent upon the physical properties of the soil, especially particle size distribution, and hydraulic forces. Without careful design, migrating soil fines may accumulate in the collector/drain or accumulate at the soil/collector interface and potentially stop the flow of groundwater, rendering the collector drains ultimately useless.

7.5.3 Seepage Control with Interceptor Drains. Seepage is the slow movement of gravitational water through soil and rock; the flowing water creates a frictional drag force proportional to flow velocity. Seepage forces have been significant contributors to the failures of dams and other earthen structures, in addition to subsurface structures such as foundations and retaining walls.

In broad terms, there are three ways to deal with seepage groundwater. In a few cases, the structure can be designed to allow passage of groundwater without endangering the integrity and performance of the structure. The remaining two seepage control methods are one that mostly keeps the water out and another that uses drainage methods to control water removal. The techniques for shielding a structure involve wall or curtain construction, as barriers are not considered drainage materials.

Controlling seepage via drainage techniques essentially involves the placement of a subsurface drain so that it intercepts the groundwater and

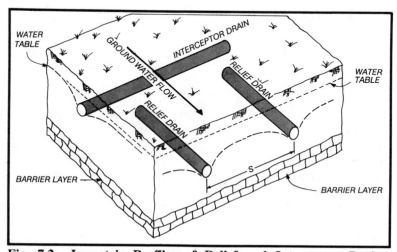

Fig. 7.2 Isometric Profiles of Relief and Interceptor Drains. Source: Ref. 3.1.21, "Drainage of Agricultural Land, U.S.D.A. SCS."

18

protects the structure that otherwise would be affected by the seepage force. The general construction of such interceptor drains is similar to the relief drains described in 7.5.2, except they commonly extend deeper into the soil profile. Good examples of the use and effectiveness of seepage interceptor drains are the long-standing practice of surrounding building foundations with highly permeable materials such as sand or aggregate. Retaining wall designs also frequently have nearly full-height drainage material to intercept seepage water.

There are many design similarities between relief drains and interceptor drains. Likewise, the precautions regarding long-term performance are similar. In both cases, extreme care is required to balance discharge capacity with inflow expectations and, furthermore, to evaluate the soil environment and inflow area materials to protect against reductions of inflow or discharge capacities.

Also, similar to subsurface relief drains, geocomposites of sufficient size and geometrical shape may be employed as interceptors, and these are particularly adaptable to rigid flat surfaces such as retaining walls and foundations.

7.5.4 Water Control of Pavement Structural Sections.
Concrete and asphalt areas (including highways, streets, parking lots, and parks) usually are supported by an underlayer of sand or stone, which should provide both structural stability and drainage for the total structure. However, significant amounts of water can enter this base or subbase course from a variety of sources and, without the opportunity to drain, remain entrapped for extended time periods. Combinations of vehicular loadings and freeze/thaw cycles on pavements with saturated base layers cause large hydrostatic forces that accelerate breakdown of the pavement. Such deterioration is widely experienced on heavily traveled roadways and is considered the major cause of deterioration on most highway systems in the United States. Other paved areas also suffer shortened lives due to inadequate subgrade drainage. Included in this group are airport runways and taxiways, residential streets, parking lots, sidewalks, and tennis courts. Any outdoor pavement should be designed for rapid removal of infiltrated water in the base course following rainfall. Federal Highway Administration (FHWA) training materials suggest that the saturation level be reduced to 85% or less within two hours after rainfall ceases.

The design of pavement drainage generally involves both the base layer and a specific drainage product or system. Designs require a drainable structural base or subbase course, employing open graded materials. Low-permeability sand bases, except for small areas with light loadings (non-vehicular), may be inadequate.

Open-graded permeable structural courses slope toward the drainage system, where water from the course is collected and discharged. Conventional roadway design is to slope the course following the roadway center crown or elevation and place a collector drain system at the downslope edge, which may be both sides of a relatively flat roadway or one side of a superelevated curve. Once the aggregate layer drains freely into the edge drain(s), the drain must function as a relief drain of significantly greater length. Because of the long runs in typical edge drain installation, side outletting into a drainage ditch or storm drain is necessary for outflow/inflow continuity. Outlet spacings are typically several hundred feet on centers for pipe or high-column edge drain systems on major highways.

7.6 Inflow-Outflow Continuity
7.6.1 Flow Continuity for Maximum Performance.
For adequate subsurface drainage system performance in any application, designs balancing outflow and inflow are essential. All sources of inflow must be identified and quantified. Inflow sources are most commonly: (a) surface infiltration (rainfall, snowfall, irrigation sources); (b) water transfer from adjacent areas (springs, waterfalls, and watersheds); and (c) high water table.

The flow of water through soil and/or pervious base material into subsurface collectors, accompanied by discharge and disposal, is dependent upon several factors: (a) soil permeability (hydraulic conductivity); (b) flow energy losses at collectors (such as siltation of filter envelopes or fabric, convergence of flowlines, entrance resistance of perforations); and (c) hydraulic characteristics of the subsurface drain system.

Soil permeability is carefully evaluated because it is a requisite element to subsurface drainage performance. Flow reduction potential due to convergence and siltation must be considered because these also contribute significantly to the result. Equally important is the determination of hydraulic properties for the collector/drain systems to permit unrestricted discharge of collected water.

7.7 Determining Inflow to Collectors
7.7.1 Relief Drains.
The USDA Soil Conservation

Service procedure for calculating the flows to relief drains involves computation of drain spacing given the hydraulic conductivity of the soil, the depth to the drain and depth to any impermeable barriers, and the desired level of water table. The spacing is computed using either the ellipse equation or the modified ellipse equation as described in Section 7.4. The area served by parallel relief drains is equal to the spacing times the length of the drain plus one-half the spacing. The discharge can be expressed by the equation:

$$Q_r = \frac{q \cdot S \cdot (L + S/2)}{43,200} \qquad \text{(Eq. 7.2, customary units)}$$

where: Q_r = relief drain discharge, cubic feet per second; q = drainage coefficient, inches per hour; S = drain spacing, feet; L = drain length, feet or

$$Q_r = q \cdot S \cdot (L + S/2) \qquad \text{(Eq. 7.2a, SI units)}$$

where: Q_r = relief drain discharge, cubic meters per day; q = drainage coefficient, meters per day; S = drain spacing, meters; and L = drain length, meters. Figure 7-3 shows a parallel relief drain system; the shaded area indicates the area served by one of the relief drains.

The value of Q_r is for the drain in its entirety, i.e., at its discharge to a collector drain. Therefore, the flow into the relief drain will be uniform along its length, with the discharge at any point along the drain proportional to the length. This suggests that collector drains receiving flow from several relief drains can be sized according to the flow carried at the various points along the collector (i.e., not uniform diameter). Depending on the size, shape, and configuration of the area to be drained, this variable sizing could result in significant cost savings.

7.7.2 Wick Drains and Chimney Drains. Groundwater moves into the drain from the surrounding soils due to higher pore pressures; soil consolidation rates are accelerated by the escape of water from the soil voids. Volume flow rate to these drains is a function of the number and type of alternating soils in the profile, the permeability and porosity of the soil, extent of soil disturbance during installation of the drains, the radial flow resistance of the drain, blinding/clogging potential of the sand or geotextile wrap, and, in the case of geocomposite drains, the amount of free (open) inflow surface, the effective size of the drain, and the number of drains installed.

7.7.3 Interceptor Drains. In calculating the flow to an interceptor drain from up-slope sources, the Darcy equation may be used. The flow per unit length into an interceptor drain may be calculated using the equation:

$$q_u = K_e \cdot i \cdot A \cdot (y/(y + d)) \qquad \text{(Eq. 7.3)}$$

where: q_u = unit flow into drain, cubic feet per second per foot (cubic meters per second per meter)

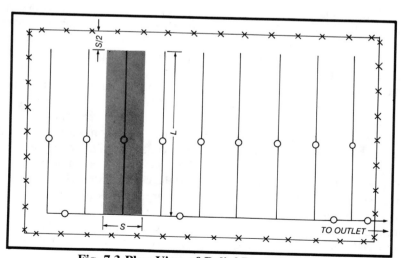

Fig. 7.3 Plan View of Relief Drain System

and K_e = effective saturated lateral hydraulic conductivity of the strata above the barrier, feet per second (meters per second), computed from:

$$K_e = \frac{K_1 T_1 + K_2 T_2 + ... + K_n T_n}{T_1 + T_2 + ... + T_n} \qquad \text{(Eq. 7.4)}$$

where: K_n = hydraulic conductivity for layers K_1 through K_n of the soil profile, feet per second (meters per second); T_n = thickness of layers T_1 through T_n, feet (meters); i = slope normal to groundwater contours, feet per foot (meter per meter); A = area through which flow occurs, square feet (square meters); y = height of maximum water surface immediately above proposed drain, feet (meters); and d = distance from drain invert to impermeable barrier, feet (meters).

Generally, the maximum water table height would be used to obtain the saturated depth from which K is obtained. This same depth would be used to obtain the area A for a unit width. The plane along which the area must be obtained is parallel to the contours or normal to the direction of flow.

7.7.4 Perimeter and Edge Drains. Perimeter and edge drains function similarly to relief drains. Used longitudinally on roads and airport runways, edge drains are designed to quickly remove excess water from the base or subbase course and subgrades of pavement structures to prevent premature deterioration of the pavement.

Flow into an edge drain or perimeter drain is similar to that of interceptor drains discussed previously, except most flow enters from the pavement side of the drain. Side outlets for edge drains must be spaced according to the estimated flow in the edge drain and the hydraulic flow capacity of the drain. In using some of the newer geocomposite fin drains, laboratory tests for inflow and capacity of each specific product being evaluated must be used for comparisons of appropriateness, since specific criteria have not been adopted by consensus.

7.8 Hydraulics of Subsurface Drains

7.8.1 Outflow of Collected Water. The final design of subsurface drainage systems concerns disposal of the water collected from the designated area. Outflow rate potential must at least equal the expected inflow rate, requiring diligent hydraulic analysis of the outlet system.

7.8.2 Fundamental Hydraulic Theory for Drains. Flow to and within subsurface drainage systems is normally gravity influenced. Discharge rates are dependent upon the slope of the drain, hydraulic characteristics of the drainage medium or material (aggregate, pipe, geocomposite), and flow energy losses inherent to the system.

The differences in elevation (slope) establish the gravity flow energy potential of the drain. Disposal water flows downhill within the drain at increasing rates with steeper slopes, although not always linearly. Flow resulting from the inclination of the drain derives increased energy from the kinetics of flow, which is nonlinear to the flow velocity.

Offsetting the energy-promoting water flow are energy losses within the drain that reduce flow. Within the drain, most energy losses relate to friction, roughness of the interfaces, and velocity fluctuations created by turbulent flow. Pipe drains (smooth or corrugated), aggregate or sand drains, composite aggregate/pipe drains, geocomposites, and geo-synthetic fabrics all have different features that contribute to flow reductions.

7.8.3 Hydraulics of Pipe Drains. The most convenient formula for evaluating flow rates in pipes subject to gravity flow and without changes in direction or size is that proposed by Manning in 1889:

$$Q = \frac{1.49\, AR^{2/3} S^{1/2}}{n} \qquad \text{(Eq. 7.5, customary units)}$$

where: Q = flow volume, cubic feet per second; A = cross-sectional area of the water stream, square feet; R = hydraulic radius; R = A/WP, feet; S = slope of the drain ($\Delta h/1$), feet per foot; WP = wetted perimeter, which is the length of the periphery of the cross-sectional shape of the liquid in contact with the pipe wall surface, feet; and n = roughness coefficient for the drain product (values for n depend on interior surface geometry and pipe material) or

$$Q = \frac{AR^{2/3} S^{1/2}}{n} \qquad \text{(Eq. 7.5a, SI units)}$$

where: Q = flow volume, cubic meters per second; A = cross-sectional area, square meters; R = hydraulic radius, meters; S = slope of the drain ($\Delta h/1$), meters per meter; WP = wetted perimeter, meters; and n = roughness coefficient.

Table 7-1.
Recommended Design Values
of Effective Roughness and Friction
Coefficients for Closed Conduits

Conduit Material	Manning "n"
Corrugated Plastic Tubing:	
a. 3–8 in. (75–200 mm) diameter	0.014–0.016
b. 10–12 in. (250–300 mm) diameter	0.016–0.018
c. Larger than 12 in. (300 mm) diameter	0.019–0.021
Concrete pipe	0.011–0.014
Corrugated Metal Pipe 1/2 x 2-2/3 in. (12.5 x 66.7 mm), corrugations, plain annular corrugations	0.022–0.026

Note: Corrugated metal pipes with helical corrugations may have lower n values under full flow conditions than shown for annular corrugated pipe.

Clay drain tile	0.011–0.014
Ductile iron pipe (cement lined)	0.011–0.014
Plastic pipe (smooth interior)	0.010–0.013
Spiral rib metal pipe	0.012–0.015

This table provides the recommended Manning's n values for estimating internal volume flow rates for the material listed. Actual pipeline performance depends upon the effects of abrasion, corrosion, deflection, alignment, joint conditions, and flow velocity.

Table 7-1 shows typical values for the roughness coefficient (n) for commonly used closed conduits.

7.8.4 Hydraulics of Geocomposite Drains.
Research on geocomposite drains is of recent vintage, and no general formula has yet been established for these products. Some general comments can be made:

(1)   Geocomposites generally have more complex flow patterns than pipes, thus contributing to higher flow energy losses.

(2)   Certain designs are more hydraulically efficient, thus providing higher flow rates per unit width or thickness.

(3)   The outer geotextile wrapping may affect hydraulic properties.

(4)   Most geocomposites react noticeably to applied earth loads.

Laboratory test procedures for geocomposites have evolved such that reliable data should be available for any product. Test data should provide flow rates in a product test situation modeling in-service environment.

7.8.5 Hydraulics of Geotextiles.
Fluid flows through geotextiles serving as a filter, with the primary function of the fabric being to restrain solids. The measure of a fabric's ability to permit water flow perpendicular to its plane is termed permittivity.

Nonwoven geotextiles are capable of transmitted fluids within the plane of the fabric. Planar flow is dependent upon the nature of the fabric, the gradient (slope), and the compressive forces on the material. Planar flow properties for any geosynthetic may be determined from ASTM D4716 procedure. Most planar flow is laminar and is expressed in gallons/ minute/foot width or cubic meters/second/meter width.

7.8.6 Hydraulics of Aggregate Drains.
In all but extremely coarse aggregates, flow velocity is low and laminar, thus permitting use of the basic Darcy equation:

$$Q = KiA \qquad \text{(Eq. 7-6)}$$

where: Q = flow rate, cubic feet per second (cubic meters per second); K = permeability of aggregate, feet per second (meters per second); i = hydraulic gradient (approximately equal to the slope ($-\Delta h/1$), feet per foot (meter per meter); A = cross-sectional area of the aggregate-filled trench, square feet (square meters).

The most significant variable is permeability, which is principally dependent upon the voids established by the aggregate's particle size and gradation. Aggregate permeability can vary widely. Laboratory testing to confirm permeability of selected aggregate versus the design requirements is highly recommended.

The amount of fines in the aggregate mix affects permeability. Generally, higher fine content results in lower permeability, indicating the desirability of clean, washed stone. Safeguards are required to minimize passage of excessive fines from adjacent soil into the aggregate drain.

7.8.7 Hydraulics of Consolidation Drains.
Flow within consolidation drains is proportional to the cross-sectional area and the hydraulic conductivity of the fill material in the case of sand drains, and flow energy is dissipated in moving the water column upward. The vertical drain must not develop significant restrictions to flow within the drain. Head losses in vertical drains are determined by:

$$h_d = \frac{q_d d}{2kA} \qquad \text{(Eq. 7-7)}$$

where: h_d = total head loss in drain, feet (meters); q_d = groundwater discharge for each drain, cubic feet per second (cubic meters per second); d = depth of the drain, feet (meters); k = permeability coefficient of drain backfill, feet per second (meters per second); and A = cross-sectional area of the drain, square feet (square meters).

Note that higher backfill permeabilities and larger cross-sectional areas reduce head losses proportionally. As consolidation progresses, the density of the backfill material may increase and there may be an accompanying decrease in permeability.

Geocomposite vertical drains may be evaluated in a fashion similar to sand drains. In determining head loss, the hydraulic transmissivity (flow rate potential) of the wick drain (determined by laboratory test) is substituted for the permeability coefficient and the cross-sectional area in Equation 7-7.

8. Structural Considerations

8.1 Loading
The loads that may be applied to an urban subsurface drainage system are categorized as either dead or live loads. Equipment live loads may be critical during construction operations.

Pipe and geocomposite systems are designed to carry live and dead loads. Geomembrane, geotextile, and aggregate systems, by their nature, normally do not have to be designed for structural strength to carry live or dead loads, but must be designed for stability and possible shear forces during construction.

8.1.1 Dead Loads may include earth loads, weight of pipe, weight of fluid in the system, building foundation loads, and surcharge loads. The primary dead load, which usually governs, is the earth load. The magnitude of the earth load is dependent on the unit weight of the soil and depth of the fill and can be determined by the design methods found in References 3.1.6 through 3.1.13.

8.1.2 Live Loads are imposed by highway vehicles, trains, aircraft, and construction equipment. Design live loads are generally taken as the HS 20 or alternate interstate loading for highways and the Cooper E 80 loading for railroads. For distribution of these loadings onto the buried structure, see Reference 3.1.13 for highway loadings, Reference 3.1.14 for railroad loadings, and Reference 3.1.15 for aircraft loadings.

8.1.3 Construction Loads from heavy construction equipment traveling over or across an installed drain system may create load concentrations in excess of design loads that may displace and damage the system. Such locations should be evaluated by the engineer to determine if displacement and damage may occur. If necessary, crossing location requirements should be detailed in the construction contract documents. A typical crossing for pipe systems consists of a temporary earth fill constructed to an elevation of at least three feet (one meter) over the top of the system and to a width sufficient to prevent lateral displacement of the system.

8.2 Embedment

The structural performance of buried pipe is dependent on the interaction between the soil and pipe. Therefore, the pipe embedment must be selected for structural and drainage characteristics. Structural characteristics of the embedment include consideration of the dimensions of the embedment around the pipe; the type of soil, density, and compaction of the embedment, native soil, and fill; the depth of burial of the pipe; and the height and characteristics of the water table. The required dimensions and type of soil, density, and compaction of the embedment are dependent on the pipe stiffness.

Flexible pipes, such as plastic and corrugated metal, use the embedment materials to transfer vertical loads in to the adjacent soil. Rigid pipe, such as concrete and clay, transfer vertical loads directly into the bedding with minimal load transfer into the adjacent soil. Therefore, the required structural characteristics of the embedment vary with the type of pipe and shall be determined in accordance with Section 8.3.

The type of equipment to be used to compact the embedment and fill should be evaluated by the engineer to determine if pipe damage and displacement may result from its use. If necessary, equipment limitations should be detailed in the construction contract documents. Heavy vibratory equipment shall not be permitted to compact fill over an installed pipe system until there is at least three feet (one meter) of earth cover over the top of the pipe.

8.3 Pipe Design

There are several pipe design methods in existence. The appropriate method depends on the pipe application or end use, the type of pipe material, and possibly the project authority or owner. Acceptable design methods for pipe projects within the scope of this practice are presented for each pipe material in Sections 8.3.1 through 8.3.4.

8.3.1 Corrugated Metal Pipe. Service load and load factor design methods and embedment requirements are presented in Section 12 of Reference 3.1.13 and in Reference 3.1.8.

8.3.2 Thermoplastic Pipe. Manufacturers' literature should be checked for recommended design methods and embedment requirements. Other sources of design information are listed in the following paragraphs.

The *Handbook of PVC Pipe: Design and Construction* (Reference 3.2.34) is a unified source for PVC pipe design and construction.

ASCE Manual No. 60, "Gravity Sanitary Sewer Design and Construction" (Reference 3.2.15), presents design methods and requirements for all thermoplastic pipes.

NCHRP Report 225, "Plastic Pipe for Subsurface Drainage of Transportation Facilities" (Reference 3.2.19), presents a state-of-the-art review of thermoplastic pipe design methods.

Section 18 of the AASHTO Standard Specifications for Highway Bridges (Reference 3.1.13) presents service load and load factor design methods and embedment requirements for some PVC and PE pipe products.

8.3.3 Precast Concrete Pipe. Service load and load factor design methods and embedment requirements are presented in Section 17 of Reference 3.1.13.

8.3.4 Clay Pipe. Service load design methods of embedment requirement are presented in Reference 3.1.9.

8.4 Other Systems

Geomembranes, geotextiles, and aggregate systems, by their nature, do not have to be designed for structural strength to carry live or dead loads. These systems, however, must be designed for stability, for possible shear forces from construction operations, and to maintain maximum flow capacities after placement.

Geomembranes and geotextiles should have sufficient strength to prevent tearing during construction operation. Geocomposites should be of such shape and have sufficient strength to prevent major deformations of the geotextile into the core that will reduce flow capacity under load. For pavement applications, geocomposite drains should have sufficient strength to withstand a long-term compressive force of 20 psi (138 kPa) normal to the plane of the drain or twice the active soil pressure at the drain depth, whichever is greater. For other applications, geocomposite drains must be evaluated against the anticipated sustained dead loads to ensure that the applied loads do not exceed the proven allowable long-term load level to prevent creep deformations of the polymer materials, which would cause reduction in flow capacity.

9. Materials

9.1 Pipe

When specified, pipe shall conform to the material and manufacturing requirements of the American Society for Testing and Materials (ASTM), American Association of State Highway and Transportation Officials (AASHTO), or American Water Works Association (AWWA) standards referenced. The ASTM and comparable AASHTO standard for a product are generally identical, however, there may be some differences, especially since AASHTO standards are normally a year behind ASTM standard revisions. For example, ASTM C444 and AASHTO M175 differ in the size and spacing of perforations. Since these standards cover only material and

manufacturing requirements, installation and performance requirements must be determined by the engineer and included in the construction contract documents. If there is a separate metric edition of a standard, its designation includes the letter M, i.e., C444M. The following lists of standards are for products commonly accepted and used on current projects; it is not the intent of these listings to restrict the use of new products that may be developed and found satisfactory.

9.1.1 Concrete Pipe. Reinforced and nonreinforced concrete pipe are used for gravity flow systems. Reinforced concrete pressure pipe and prestressed concrete pressure pipes are used for pressure flow systems. Concrete fittings and appurtenances such as wyes, tees, and manhole sections are generally available. A number of jointing methods are available depending on the tightness required and the operating pressure.

A number of mechanical processes are used in the manufacture of concrete pipe. These processes use various techniques including centrifugation, vibration, packing, and tamping for consolidating the concrete in forms. Gravity and pressure concrete pipe may be manufactured to any reasonable strength requirement by varying the wall thickness, concrete strength, quantity, and configuration of reinforcing steel or prestressing elements.

Gravity Flow Applications. Concrete pipe is specified by nominal diameter, type of joint, and D-load strength or reinforcement requirements. The product should be manufactured in accordance with one or more of the following standard specifications:

ASTM C14/AASHTO M86 (ASTM C14M/ AASHTO M86M): Concrete Sewer, Storm Drain and Culvert Pipe — covers nonreinforced concrete pipe from 4 through 36-inch (100- through 900-mm) diameters in Class 1, 2, and 3 strengths.

ASTM C76/AASHTO M170 (ASTM C76M/ AASHTO M170M): Reinforced Concrete Culvert, Storm Drain, and Sewer Pipe — covers reinforced concrete pipe in five standard strengths. Class I-60 through 144-inch (1500- through 3600-mm) diameters; Class II, III, IV, and V; 12- through 144-inch (300- through 3600-mm) diameters. Larger sizes and higher classes are available as special designs.

ASTM C118 (ASTM C118M): Concrete Pipe for Irrigation or Drainage — covers concrete pipe intended to be used for the conveyance of water under low hydrostatic heads, generally not exceeding 25 feet (75 kPa), and for drainage in sizes from 4- through 24-inch (100- through 600-mm) diameters in standard and heavy-duty strengths.

ASTM C361 (ASTM C361M): Reinforced Concrete Low-Head Pressure Pipe — covers reinforced concrete pipe conduits with low internal hydrostatic heads generally not exceeding 125 feet (375 kPa) in sizes from 12- through 108-inch (100- through 2700-mm) diameters.

ASTM C412/AASHTO M178 (ASTM C412M/ AASHTO M178M): Concrete Drain Tile — covers nonreinforced concrete drain tile with internal diameters from 4- through 24-inch (100- through 600-mm) for standard quality, and 4- through 36-inch (100- through 900-mm) for extra-quality, heavy-duty extra- quality, and special quality concrete drain tile.

ASTM C444/AASHTO M175 (ASTM C444M/AASHTO M175M): Perforated Concrete Pipe — covers perforated concrete pipe intended to be used for underdrainage in 4-inch (100-mm) and larger diameters.

ASTM C478/AASHTO M199 (ASTM C478M/AASHTO M199M): Precast Reinforced Concrete Manhole Sections — covers precast reinforced-concrete manhole risers, grade rings, and tops to be used to construct manholes for storm and sanitary sewers.

ASTM C505 (ASTM C505M): Nonreinforced Concrete Irrigation Pipe with Rubber Gasket Joints — covers pipe to be used for the conveyance of water with working pressures, including hydraulic transients, of up to 30 feet (90 kPa) of head. Higher pressures may be used up to a maximum of 50 feet (150 kPa) for 6- through 12-inch (150- through 300-mm) diameters, and 40 feet (120 kPA) for 15- through 18-inch (375- through 450-mm) diameters by increasing the strength of the pipe.

ASTM C506/AASHTO M206 (ASTM C506M/AASHTO M206M): Reinforced Concrete Arch Culvert, Storm Drain, and Sewer Pipe — covers reinforced-concrete arch pipe in sizes from 15- through 132-inch (375- through 3300-mm) equivalent circular diameters. Larger sizes are available as special designs.

ASTM C507/AASHTO M207 (ASTM C507M/AASHTO M207M): Reinforced Concrete Elliptical Culvert, Storm Drain, and Sewer Pipe — covers reinforced elliptical concrete pipe in five standard classes of horizontal elliptical, 18 through 144 inch (450 through 3600mm) in equivalent circular diameter and five standard classes of vertical elliptical, 36 through 144 inches (900 through 3600 mm) in equivalent circular diameter are included. Larger sizes are available as special designs.

ASTM C654/AASHTO M176 (ASTM C654M/AASHTO M176M): Porous Concrete Pipe — covers porous nonreinforced-concrete pipe in sizes from 4- through 24-inch (100- through 600-mm) diameters and in two strength classes.

ASTM C655/AASHTO M242 (ASTM C655M/AASHTO M242M): Reinforced Concrete D-Load Culvert, Storm Drain and Sewer Pipe — covers acceptance of pipe design and production based upon the D-load concept and statistical sampling techniques.

ASTM C789/AASHTO M259 (ASTM C789M/AASHTO M259M): Precast Reinforced Concrete Box Sections for Culverts, Storm Drains, and Sewers — covers box sections with 2 or more feet (0.6 m) of earth cover when subjected to highway live loads, and zero cover or greater when subjected to only dead load in sizes from 3-foot (900-mm) span by 2-foot (600-mm) rise to 12-foot (3600-mm) span by 12-foot (3600-mm) rise.

ASTM C850/AASHTO M273 (ASTM C850M/AASHTO M273M): Precast Reinforced Concrete Box Sections for Culverts, Storm Drains, and Sewers with Less than 2 Feet (0.6 m) of Cover Subject to Highway Loading — covers box sections with less than 2 feet (0.6 m) of earth cover in sizes from 3-foot (900-mm) span by 2-foot (600-mm) rise to 12-foot (3600-mm) span by 12-foot (3600-mm) rise.

ASTM C985 (ASTM C985M): Nonreinforced Concrete Specified Strength Culvert, Storm Drain, and Sewer Pipe — covers acceptance of nonreinforced concrete pipe design and production based on specified strengths and statistical sampling techniques.

Pressure Flow Applications. Concrete pressure pipe is specified to provide custom designs based on specific conditions of service. The product should be designed and manufactured in accordance with one or more of the following standards.

AWWA C300: Reinforced Concrete Cylinder Pipe — covers reinforced-concrete cylinder pipe from 24- through 144-inch (600- through 3600-mm) diameters and larger, and standard lengths in the 12- through 24-foot (3.6- through 7.2-m) range. Although the maximum loads and pressures for this type of pipe depend on the pipe diameter, wall thickness, and strength limitations of the concrete and steel, it is uncommon for this pipe to be used at pressures over 250 psi (1750 kPa) or in trench installations exceeding 20 feet (6 m) of earth cover.

AWWA C301: Prestressed Concrete Cylinder Pipe — covers prestressed concrete cylinder pipe from 24- through 144-inch (600- through 3600-mm) diameters and larger. Pipe larger than 250 inches (6250 mm) in diameter have been produced. Standard lengths in the 16- to 24-foot range (4.8- to 7.2-m), and longer lengths, are available. Prestressed concrete cylinder pipe has been designed for operating pressures over 400 psi (2800 kPa) and earth covers in excess of 100 feet (30 m).

AWWA C302: Reinforced Concrete Noncylinder Pipe — covers reinforced-concrete noncylinder pipe from 12- through 144-inch (300- through 3600-mm) diameters and larger. Reinforced concrete noncylinder pipe, because it does not contain a watertight membrane (steel cylinder), is limited to internal pressures of 55 psi (385 kPa) or less.

AWWA C303: Pretensioned Concrete Cylinder Pipe — covers pretensioned concrete cylinder pipe from 10- through 48-inch (250- through 1200-mm) diameters and larger. Standard lengths are generally 36 to 40 feet (10.8 to 12 m). Pretensioned concrete cylinder pipe has been designed for pressures over 400 psi (2800 kPa).

9.1.2 Thermoplastic Pipe. Thermoplastic pipe materials include a broad variety of plastics that can be repeatedly softened by heating and hardened by cooling through a temperature range characteristic for each specific plastic, and in the softened state can be shaped by molding or extrusion. Generally, thermoplastic pipe materials are limited to acrylonitrile-butadiene-styrene (ABS), polyethylene (PE), and polyvinyl chloride (PVC). Thermoplastic pipes are produced in a variety of shapes and dimensions. Pipe properties can be modified by changing the wall thickness or profile, for both pressure and non-pressure applications.

9.1.2.1 Acrylonitrile-Butadiene-Styrene (ABS) Pipe. ABS pipe is manufactured by extrusion of ABS material and is limited to gravity flow applications. ABS composite pipe is manufactured by extrusion of ABS material with a series of truss annuli that are filled with filler material such as lightweight Portland cement concrete. ABS fittings are available for the product. The jointing systems available include elastomeric gasket joints and solvent cement joints.

Gravity Flow Applications. ABS pipe should be manufactured in accordance with one of the following standard specifications.

ASTM D2680/AASHTO M264: Acrylonitrile-Butadiene-Styrene (ABS) and Poly (Vinyl Chloride) (PVC) Composite Sewer Piping — covers ABS or PVC composite pipe, fittings, and a joining system for nonpressure systems.

ASTM D2751: Acrylonitrile-Butadiene-Styrene (ABS) Sewer Pipe and Fittings — covers ABS pipe and fittings from 3- through 12-inch (75- through 300-mm) diameter.

9.1.2.2 Polyethylene (PE) Pipe. PE pipe is used for both gravity and pressure flow systems. PE pipe is manufactured by extrusion of PE plastic material. PE pipe is specified by material designation, nominal diameter (inside or outside), standard dimension ratios, ring stiffness, and type of joint. PE fittings are available.

Gravity Flow Applications. PE pipe for gravity flow applications should be manufactured in accordance with one or more of the following standard specifications.

ASTM F405/AASHTO M252: Corrugated Polyethylene Tubing and Fittings — ASTM F405 covers pipe from 3- through 6-inch (75- through 150-mm) diameters. AASHTO M252 covers pipe from 3- through 10-inch (75- through 250-mm) diameters.

ASTM F667/AASHTO M294: Large Diameter Corrugated Polyethylene Tubing and Fittings — ASTM F667 covers pipe from 8- through 24-inch (200- through 600-mm) diameters. AASHTO M294 covers pipe from 12- through 36-inch (300- through 900-mm) diameters.

ASTM F810: Smoothwall Polyethylene (PE) Pipe for Use in Drainage and Waste Disposal Absorption Fields — covers smoothwall PE pipe, including coextruded, perforated, and nonperforated from 3- through 6-inch (75- through 150-mm) diameters.

ASTM F892: Polyethylene (PE) Corrugated Pipe with a Smooth Interior and Fittings — covers corrugated PE pipe 4 inches (100 mm) in diameter.

ASTM F894: Polyethylene (PE) Large Diameter Profile Wall Sewer and Drain Pipe — covers profile wall PE pipe from 18- to 120-inch (450- to 3000-mm) diameters for both low-pressure and gravity flow applications.

Pressure Flow Applications. PE pressure pipe should be manufactured in accordance with one or more of the following standard specifications.

ASTM D2239: Polyethylene (PE) Plastic Pipe (SIDR-PR) Based on Controlled Inside Diameter — covers PE pipe with 0.5- to 6-inch (12- to 150-mm) diameters and with pressure ratings from 400 to 800 psi (2800 to 5600 kPa).

ASTM D3035: Polyethylene (PE) Plastic Pipe (SDR-PR) Based on Controlled Outside Diameter — covers PE pipe 0.5- to 6-inch (12- to 150-mm) diameters and with pressure ratings from 400 to 800 psi (2800 to 5600 kPa).

9.1.2.3 Polyvinyl Chloride (PVC) Pipe. PVC pipe is used for both gravity and pressure flow systems. PVC pipe is manufactured by extrusion of the material. PVC composite pipe is manufactured by extrusion of PVC material with a series of truss annuli that are filled with material such as lightweight Portland cement concrete. PVC pipe is specified by nominal diameter, dimension ratio, pipe stiffness, and type of joint. PVC pressure and nonpressure fittings are available.

Gravity Flow Applications. PVC pipe for gravity flow applications should be manufactured in accordance with one or more of the following standard specifications.

ASTM D2680/AASHTO M264: Acrylonitrile-Butadiene-Styrene (ABS) and Poly (Vinyl Chloride) (PVC) Composite Sewer Piping — covers ABS or PVC composite pipe, fittings, and a joining system for nonpressure sanitary sewer and storm drain systems in 6- through 15-inch (150- through 375-mm) diameters.

ASTM D2729: Poly (Vinyl Chloride) (PVC) Sewer Pipe and Fittings — covers material and test requirements for PVC pipe and fittings for sewer and drain pipe in sizes from 2- to 6-inch (50- to 150-mm) diameters. Standard perforations available only in 4-inch (100-mm) diameter pipe.

ASTM D3034: Type PSM Poly (Vinyl Chloride) (PVC) Sewer Pipe and Fittings — covers material and test requirements for PVC pipe and fittings for sewer pipe systems from 4- through 15-inch (100- to 375-mm) diameters.

ASTM F679: Poly (Vinyl Chloride) (PVC) Large-Diameter Plastic Gravity Sewer Pipe and Fittings — covers material and test requirements for PVC gravity sewer pipe and fittings from 18- through 27-inch (450- through 675-mm) diameters, with integral bell elastomeric seal joints and smooth inner walls.

ASTM F758: Smooth-Wall Poly (Vinyl Chloride) (PVC) Plastic Underdrain Systems for Highways, Airport, and Similar Drainage — covers material and test requirements for smooth wall pipe and fittings for PVC underdrains from 4- through 8-inch (100- through 200-mm) diameters with perforated or nonperforated walls for use in subsurface drainage systems.

ASTM F789: Type PS-46 Poly (Vinyl Chloride) (PVC) Plastic Gravity Flow Sewer Pipe and Fittings — covers materials and test requirements for PVC gravity sewer pipe and fittings from 4- through 18-inch (100- through 450-mm) diameters, with a minimum pipe stiffness of 46 psi (320 kPa).

ASTM F794: Poly (Vinyl Chloride) (PVC) Large-Diameter Ribbed Gravity Sewer Pipe and Fittings Based on Controlled Inside Diameter — covers materials and test requirements for PVC gravity sewer ribbed pipe and fittings from 4- through 48-inch (200- through 1200-mm) inside diameters, with integral bell and elastomeric seal joints.

ASTM F800: Corrugated Poly (Vinyl Chloride) (PVC) Tubing and Compatible Fittings — covers materials and test requirements for perforated and nonperforated PVC tubing and fittings from 4- through 12-inch (100- through 300-mm) diameters for use in agricultural and other soil drainage and septic tank effluent beds.

ASTM F949: Poly (Vinyl Chloride) (PVC) Corrugated Sewer Pipe with a Smooth Interior and Fittings — covers materials and test requirements for

PVC pipe and fittings from 4- to 10-inch (100- through 250-mm) diameters with corrugated outer wall fused to a smooth inner wall for sanitary and storm sewers and perforated and nonperforated pipe for subdrainage.

Pressure Flow Applications. PVC pressure pipe should be manufactured in accordance with one of the following standard specifications.

ASTM D1785: Poly (Vinyl Chloride) (PVC) Plastic Pipe, Schedules 40, 80, 120 — covers materials and test requirements for PVC pipe pressure rated for water transmission in diameters ranging from 1/8- through 12-inch (3- through 300-mm) diameters.

ASTM D2241: Poly (Vinyl Chloride) (PVC) Pressure Rated Pipe (SDR Series) — covers materials and test requirements for PVC pipe pressure rated for water transmission in sizes ranging from 1/8- through 36-inch (3- through 900-mm) outside diameters. Available with SDRs ranging from 13.5 to 64, depending on the diameter.

AWWA C900: Poly (Vinyl Chloride) (PVC) Pressure Pipe, 4 in. (100-mm) through 12 in. (300-mm), for Water — covers materials and test requirements for PVC pipe pressure rated for water transmission in sizes ranging from 4- through 12-inch (100- through 300-mm) diameters, and three pressure classes of 100, 150, and 200 psi (700, 1050, and 1400 kPa) with DRs of 25, 18, and 14, respectively, and providing a hydrostatic design basis (HDB) of 4000 psi (28 MPa).

AWWA C905: Poly (Vinyl Chloride) (PVC) Water Transmission Pipe, Nominal Diameters 14 in. (350 mm) through 36 in. (900 mm) — covers materials and test requirements for PVC pipe pressure rated for water transmission in sizes ranging from 14- through 36-inch (350- through 900 mm) outside diameters, and DRs of 18, 21, 25, 26, 32.5, and 41, and providing a hydrostatic design basis (HDB) of 4000 psi (28 MPa) for pressure ratings ranging from 100 to 235 psi (700 to 1650Pa).

9.1.3 Metal Pipe. Corrugated metal pipe is fabricated from corrugated steel or aluminum sheets or coils. Corrugated metal pipe is specified by size, shape, wall profile, gauge or wall thickness, and coating or lining. Appurtenances, including tees, wyes, elbows, and manholes, are available. Corrugated metal pipe is limited to gravity flow applications.

Gravity Flow Applications. Corrugated metal pipe should be manufactured in accordance with one or more of the following standard specifications.

ASTM A760/AASHTO M36: Corrugated Steel Pipe, Metallic-Coated for Sewers and Drains — covers metallic-coated corrugated steel pipe from 4- through 144-inch (100- through 3600-mm) diameters.

ASTM A762/AASHTO M245: Corrugated Steel Pipe, Polymer Precoated for Sewers and Drains — covers polymer precoated corrugated steel pipe from 4- through 144-inch (100- through 3600-mm) diameters.

ASTM A849/AASHTO M190: Post Coated (Bituminous) Corrugated Steel Sewer and Drainage Pipe — covers post coated corrugated steel pipe from 4- through 144-inch (100- through 3600-mm) diameters.

ASTM B745/AASHTO M196: Corrugated Aluminum Pipe for Sewers and Drains — covers corrugated aluminum pipe from 4- through 144-inch (100- through 3600-mm) diameters.

9.1.4 Vitrified Clay Pipe (VCP). VCP is manufactured from clays and shales and vitrified at high temperatures. VCP is available in standard and extra-strength classifications, and specified by nominal pipe diameter, strength, and type of joint. The product is limited to gravity flow applications.

Gravity Flow Applications. The product should be manufactured in accordance with one or more of the following standard specifications.

ASTM C4/AASHTO M179: Clay Drain Tile — covers drain tile from 4- through 30-inch (100- through 750-mm) diameters in standard, extra-quality, and heavy-duty strengths.

ASTM C498: Clay Drain Tile, Perforated — covers perforated drain tile from 4- through 18-inch (100- through 450-mm) diameters in strengths of standard, extra-quality, heavy-duty, and extra-strength.

ASTM C700/M65: Clay Pipe, Vitrified, Extra-Strength, Standard Strength, and Perforated — covers perforated and nonperforated pipe from 3- through 42-inch (75- through 1050-mm) diameters in extra-strength and standard strength.

9.2 Other Materials and Products

Geocomposites, geomembranes, geotextiles, aggregates, wick drains, and pump and lift stations are not covered by national standard specifications. The requirements for such materials and products must be specified in construction contract documents by the engineer.

10. Codes and Permits

10.1 General

In the concept stages of an urban subsurface drainage project, preferably before the site analysis and system configuration phases, it is important to obtain copies and an understanding of all applicable federal, state, and local codes. At the same time, all federal, state, and local permits for the project should be identified, and the requirements and submittal timing of each clearly understood.

10.2 Codes

Federal, state, and local codes that apply to the design and operation of an urban subsurface drainage system shall be considered.

Federal, state, and local codes that apply to the construction of an urban subsurface drainage system shall be considered.

10.3 Permits

Federal, state, and local temporary permits that are necessary prior to and during construction of an urban subsurface drainage system project shall be secured by the owner, owner's agent, or contractor prior to construction. Any permanent permits, such as the Corps of Engineers' 404 Permit, which must be maintained after construction of the project, shall be secured by the owner and/or the owner's agent.

A copy of any temporary and permanent permits secured by the owner, owner's agent, and/or contractor shall be included as part of the contract documents. Copies of all permits secured should be furnished to the owner.

The contractor is responsible to conform to the terms and provisions of all permits during construction as stated in the contract documents.

Standard Guidelines for
Installation of Urban Subsurface Drainage

Contents

FOREWORD

The "Standard Guidelines for Installation of Urban Subsurface Drainage" is an independent document intended to complement the ASCE Urban Subsurface Drainage Manual. This guideline is a companion to the Standard Guidelines for Design of Urban Subsurface Drainage and the Standard Guidelines for Operation and Maintenance of Urban Subsurface Drainage. These standard guidelines were developed by the Urban Subsurface Drainage Standards committee, which is responsible to the Special Standards Division of the Management Group "F" (MGF) on Codes and Standards.

The material presented in this publication has been prepared in accordance with recognized engineering principles. This standard guidelines should only be used by or under the direction of an engineer competent in the field of subsurface drainage. The publication of the material contained herein is not intended as a representation or warranty on the part of the American Society of Civil Engineers, or of any other person named herein, that this information is suitable for any general or particular use, or promises freedom from infringement of any patent or patents. Anyone making use of this information assumes all liability from such use.

Standard Guidelines for
Installation of Urban Subsurface Drainage

1.0 Scope

The intent of this standard is to present installation and construction guidance for urban subsurface drainage. The collection and conveyance of subsurface drainage waters is within the purview of this standard for applications such as airports, roads, and other transportation systems, and industrial, residential, and recreation areas. Incidental surface water is considered. This document is intended for guidance in the installation phase for the engineer, project manager, inspector, and contractor.

Some of the applications this standard does not address include agricultural drainage, landfills, recharge systems, storm sewer systems, and injection systems.

2.0 Definitions

2.1 General

This section defines specific terms for use in these guidelines. References included in Section 3.0 may also be helpful in understanding the terms of these guidelines.

2.2 Terms

Envelope — permeable material placed around a pipe or drainage product to improve flow and soil filtration conditions in the area immediately adjacent to the drain, and for improving bedding and structural backfill conditions.

Geocomposite — geosynthetic materials for collecting and transporting water while maintaining soil stability.

Geomembrane — sheet materials intended to form an impermeable barrier.

Geosynthetic — a generic term for synthetic material or structure used as an integral part of a project, structure, or system. Within this category are subsurface drainage and water control products such as geomembranes, geotextiles, and geocomposites.

Geotextile — a woven or nonwoven thermoplastic sheet material intended to allow the passage of water, but not fines, and without collecting fines at the soil-textile interface.

Grade — may refer to either (1) the slope of the drain in vertical units/horizontal units; or (2) the specified vertical location of the drain; depending on the context in which it is used.

OSHA — Occupational Safety and Health Administration, the federal agency responsible for safety and health concerns on construction job sites.

Subsurface water — all water beneath the ground or pavement surface. Sometimes referred to as groundwater.

3.0 References

The references for the Standard Guidelines for Installation of Urban Subsurface Drainage are the same as the references for the Standard Guidelines for the Design of Urban Subsurface Drainage.

4.0 Contract Documents

4.1 General

Plans and Specifications typically make up the Construction Contract Documents. Other documents become part of the contract by reference.

4.2 Plans

Plans are drawings, prepared to a scale, showing the proposed subsurface drainage system and known surface and subsurface features that may affect the new installation and construction. The plans typically show type, size, material, grade, and location of the subsurface drainage system.

4.3 Specifications

Specifications are written text and/or details that provide specifics of the construction. They typically detail all appropriate standards required for the product and project. Construction methods may be specified on a case-by-case basis.

4.4 Other

Many other publications are made part of the contract documents by reference only, such as government requirements, permits, reports, or trade and industry specifications. Some of these reports will include soil borings, past construction observations, and studies.

5.0 Site Inspection

5.1 General

It is necessary to examine the plans and specifications and make a personal examination of the site and its surroundings prior to construction. This inspection should include reviews of both surface and subsurface features.

5.2 Surface Features

The surface features of the site should be located through a topographic survey and shown on the plans. The plans should be compared with existing field conditions to determine whether there are any differences between the topographic survey and present conditions. Discrepancies are to be brought to the attention of the engineer or project manager.

5.3 Subsurface Features

Subsurface features principally consist of utilities and geological conditions. All subsurface conditions are subject to field verification by the contractor prior to construction.

5.3.1 Utilities. The location and size of sanitary sewers, drains, culverts, gas lines, water mains, electric lines, telephone conduits, and other underground utilities and structures should be shown on the plans. This information should be obtained from both field surveys and other available records. The contractor should verify the information given on all existing utilities affecting work prior to construction.

5.3.2 Geologic Conditions. All appropriate and available geological conditions should be shown on the plans. An assessment should be made with respect to rock and groundwater conditions.

6.0 Installation

6.1 General

Prior to construction, all documents, including plans and specifications, subsurface information,

standard details, product shop drawings, and special provisions should be reviewed and any questions resolved.

6.2 Safety

The contractor is responsible for construction site safety. Federal regulations covering safety for all types of construction are published in the Safety and Health Regulations for Construction under the Department of Labor, Occupational Safety and Health Administration (OSHA). Many states, municipalities, and other local agencies have established codes of safe practice regarding construction. These regulations apply to all types of construction, including alteration and repair work. All personnel associated with the construction should be familiar with the requirements applicable to subsurface drainage system projects, especially in regard to safe trenching procedures.

6.3 Soil Erosion and Sediment Control

Erosion and sediment control at the site should be in accordance with federal, state, municipal, and local agency regulations, or as otherwise established by the contract plans and specifications.

6.4 Site Preparation

Site preparation should be in accordance with the contract plans and specifications and may include topsoil stripping, clearing and grubbing, pavement and sidewalk removal, rough grading, protection or relocation of existing natural drainage, removal of unsuitable soil material, construction of access roads, detours, and protection or relocation of existing structures and utilities.

6.5 Materials Receiving, Handling, and Storage

The contractor is responsible for receiving, proper handling, and storing of all construction materials for the project. Materials damaged in shipment or at the site, which cannot be repaired, should be marked or tagged and removed from the site. All materials should be unloaded and handled with reasonable care.

Stockpiling of materials should be as near as possible to where they will be installed. All materials should be stored as recommended by the manufacturer. All materials should be stockpiled in a safe manner.

6.6 Line and Grade

The contractor is responsible for maintaining all line and grade, monuments, control points, and stakes set by the engineer or surveyor until the project is completed and accepted.

For all subsurface drainage systems, line and grade should be in accordance with the contract plans and specifications.

All subsurface drainage systems should be installed true to line and grade in accordance with the contract plans and specifications. Adjustments to correct departures from specified line and grade shall not exceed those permitted by contract documents or manufacturers' recommendations, whichever are more restrictive, provided that such corrections never result in a reversal of the slope in the drainage system. Moreover, realignments must never result in buckling or other deformations of flexible drainage materials, such as flexible pipes and geocomposites, that damage or reduce the flow capacity of the system. The return to specified line and grade should be made by adding or removing bedding material, and the use of wedges or blocks is unacceptable.

The maximum allowable departure from the horizontal alignment should be specified by the contract documents. Departure distances must be determined by measurement along common elements of the planned versus actual installation, such as centerline-to-centerline measurements.

The relatively complex hydraulic characteristics of geocomposite drainage systems requires that strict controls be placed on departures from specified vertical alignment (grade). When departures occur, the return to specified grade should be at a rate not exceeding 15% of the specified grade or at a rate established by the manufacturer, whichever is more restrictive.

6.7 Excavation

Excavation should be in accordance with contract plans and specifications and may include trenching, backfilling, embankment construction, soil stabilization, and control of groundwater and surface drainage. Adequate knowledge of subsurface conditions is required for all types of excavation. Additional exploration and analysis are recommended if the subsurface information on the plans is insufficient. The contractor should notify the owner if archaeological items are encountered during construction.

6.7.1 Excavation Limits. Excavation, installation, and backfill operations should be performed in a timely matter to reduce open trench time. The length of open trench should comply with limits established by OSHA or applicable state and local regulations. Specified trench width requirements for flexible conduits should also be maintained to assure proper deflection control. Specified trench depth and width should be in accordance with the contract plans and specifications. If the trench width becomes greater than specified, the contractor should contact the engineer for a reevaluation of the required pipe strength, bedding materials and methods, and backfilling procedures to be used.

6.7.2 Handling of Excavated Material. Excavated material to be used as backfill should be stockpiled in accordance with the contract plans and specifications and applicable safety regulations a safe distance back from the edge of the trench. Generally, if trench walls are unsupported, the minimum distance from the trench side to the excavated material should be either three feet (one meter) or one-half of the trench depth, whichever is greater. If the trench walls are supported, the usual minimum distance from the trench side to the excavated material should be three feet (one meter).

These general recommendations should not supersede job-specific requirements in the presence of unstable soils and/or the potential for accumulation of water in the trenches.

For trencher installations where drainage material and backfill are placed simultaneously and where personnel are not allowed in the trench, soil placement may be unrestricted. Unused excavated materials should be disposed of in accordance with contract documents.

6.7.3 Sheathing and Shoring. OSHA and many states, municipalities, and other local agencies have established codes of safe practice regarding support requirements for trench excavation. When required by established codes, sheathing and shoring must be adequate to prevent cave-in of the trench walls or subsidence of areas adjacent to the trench, and to prevent sloughing of the base of the excavation. Sheathing and shoring should not extend into the soil envelope zone of the pipe or geocomposite drainage system. Any sheathing placed below the top of the drainage product of flexible pipes or geocomposite materials or the springline of rigid pipes should remain in place after backfilling. Movement of shoring following backfill placement may reduce the structure integrity of the surrounding embedment material.

The contractor is responsible for adequacy of any required sheathing and shoring. The strength of support systems should be based on the principles of geotechnical and structural engineering as applicable to the materials encountered.

6.7.4 Dewatering. When necessary, all excavations should be dewatered prior to and during installation and backfilling of the subsurface drain system. The contractor is responsible for dewatering operations and should ensure that foundation and bedding materials are not being removed through the dewatering system and that property damage does not result.

6.8 Foundation Preparation

Bedding for subsurface drainage systems should be as specified and completed to design line and grade. The intrusion of foreign material into any portion of the drainage system due to construction and weather events should be prevented until the system is adequately protected by backfill.

6.9 Placement of Drainage Materials

6.9.1 Pipe. A bedding material should be placed on the foundation, the pipe laid and connected, and backfill placed, all in accordance with the contract plans and specifications.

6.9.2 Geocomposite Drainage Materials. Prefabricated geocomposite subsurface collector-drains may be placed in trenches by hand or machines. Installation techniques must not cause damage to the interior core and geotextile overwrap, including any factory-made seams and connections. Joints should be made using connectors recommended by the manufacturer or in accordance with the contract plans and specifications. All joints should be made soil-tight using tape, glue, or other sealing procedures recommended by the manufacturer or in accordance with the contract plans and specifications. In all cases, such sealing procedures should provide assured long-term resistance to degradation in wet subsurface environments.

6.9.3 Other Drainage Materials. Geomembranes, geotextiles, aggregates, and pump and lift stations should be installed in accordance with the contract plans and specifications.

6.10 Backfill

Backfill material should be placed and compacted in accordance with the contract plans and specifications. Backfill should not be dumped or dropped directly on any portion of the drainage system. Heavy equipment operations should be controlled so as not to damage any portion of the drainage system. Backfill material should be placed in layers in accordance with contract documents and compacted at or near optimum moisture content.

6.11 Site Restoration

Restoration of grass, shrubs, and other plantings should be performed in accordance with contract documents. Until revegetation is complete, adequate protection against erosion and runoff is necessary, and should be in accordance with the contract documents and governing regulations. All revegetation and tree repair should be in accordance with accepted horticultural practice.

When replacing permanent pavement, the subgrade must be restored and compacted until smooth and unyielding. Thickness and type of pavement shall be as established by contract documents.

7.0 Inspection

7.1 General

The duty of the inspector is to inspect the materials furnished and the work performed to verify full compliance with the contract documents. A dedicated, qualified inspector should be on site to observe all phases of the site preparation, materials receiving and handling, installation, and site restoration. Observations of materials, workmanship, and, where specified in the contract documents, methods and means of performing construction, are required to determine compliance with contract documents.

The inspector should have unrestricted access to all areas where the preparation of the materials and parts of work to be done are carried out and conducted. The contractor shall provide access to all facilities and assistance required to perform the inspection.

7.2 Inspection of Materials

All construction materials must be carefully and thoroughly inspected prior to and during placement. Inspection should be an ongoing process, since satisfactory materials first arriving on site can be damaged during handling, storage, and installation. No material of any kind should be used prior to inspection and formal approval.

Project specifications and other product-specified information should form the basis for determining the suitability of all materials. Should any doubt concerning suitability arise, the product manufacturer should be consulted.

Shipments of select fill materials and drainage products should be accompanied by certified test reports. If such data are missing, laboratory tests should be used to confirm appropriate properties. All drainage products or drainage system components should be measured to check size, shape, and fit.

All materials must be inspected to ensure that they are free of foreign deposits, defects, and damage. Cleaning and removal of foreign matter may be acceptable, providing there is complete assurance that the construction material or product is unharmed and in "like new" condition.

Damaged products or components should be immediately removed from the site. Repairs may be performed on damaged goods following inspection and approval of the inspector and after consultation with the manufacturer. The return of any previously rejected materials, products, or components to the site is acceptable only after reinspection and approval following rework. Any material or workmanship found at any time during the construction cycle not in accordance with project specifications, for any reason, shall immediately be remedied.

7.2.1 Prefabricated and/or Premanufactured Components. Prior to installation, all prefabricated and/or premanufactured components shall be inspected to establish conformity with the project specifications and to check for damage and the presence of foreign matter. The manufacturer's certificate of compliance and product drawings should confirm compliance with the contract documents.

7.2.2 Bedding, Backfill, and Envelope Materials. The inspection of all materials for use in bedding, backfilling, and envelope materials, or as otherwise used in subsurface drainage systems, shall be checked for conformance to project specifications. The supplier's certificate of compliance shall form the basis of the inspection.

Geosynthetic products should be inspected for damage and conformity to the project specifications and drawings. The manufacturer's materials

certificate should be the basic inspection document. Any deviations from the contract specifications should be immediately referred to the project engineer.

7.2.3 Storage of Materials. Storage of materials should be managed by the contractor to avoid impairing the usability and quality of on-site materials. Observance of any special handling methods required shall be verified and recorded. Storage or special protection required by the contract documents for certain items should also be verified. The inspector is responsible for monitoring the contractor's observance of these requirements.

7.3 Inspection of Equipment

All equipment used in the construction of the project should be subject to inspection (and testing, where applicable).

7.3.1 Safety. All safety rules established by the contractor must be obeyed. Observations should be performed in a manner that will not unreasonably impede or obstruct the contractor's operations.

7.3.2 Suitability and/or Conformance. Methods and means of construction are left to the option of the contractor on most items to allow flexibility, but may be specified on items where methods and means are critical to obtaining a final desired product. While means and methods may not be specified in full detail, minimum elements are frequently identified. Where methods and means are specified in the contract documents, verification is required for compliance.

7.4 Inspection of Construction

7.4.1 General. The sequence of construction operations is an important consideration in projects that require construction in a particular order. Sequencing may also be required in order to allow the existing facilities to remain in operation during the construction phase. The contract documents normally allow as much flexibility as possible in sequencing operations and may have no requirements other than an overall completion date.

It is not the function of the inspector to supervise or direct the manner in which the work is performed.

The inspector should follow each stage of construction so that any construction errors can be resolved during construction rather than after. The inspector should closely monitor all acceptance testing for the correct test procedure. Inspection activities associated with each construction stage are summarized in Table 7-1.

7.4.2 Construction Layout. Survey controls should be established as referenced on the contract drawings. It is the contractor's responsibility to stake and build the project from the controls. All necessary auxiliary staking must be in place prior to construction. If an error in auxiliary staking is observed, detected, or suspected, the error shall be promptly called to the contractor's attention.

It is the responsibility of the inspector to ascertain that the survey control points are in place as referenced. If there is any evidence that the control points have been disturbed, the inspector shall notify the engineer, who will arrange to have the points checked by the surveyor who originally set the points.

7.4.3 Excavation and Dewatering. The inspector should confirm that all excavations and dewatering activities are performed in accordance with the contract documents and that these activities will allow construction to be completed according to plan. Proper disposal of water is the responsibility of the contractor.

Trenches shall be excavated to depths and widths as specified for correct backfill and/or envelope placement and compaction. Water standing or flowing into the trench should be removed until backfill and envelope materials are placed at the direction of the engineer.

Prior to placement of drainage products and envelope materials, all finished excavations should be inspected to ensure the absence of unsuitable materials.

7.4.4 Installation. Where the contract documents reference installation in accordance with the manufacturer's directions, such directions shall be provided for use in verifying that subsequent installation is performed in accordance with them.

7.4.5 Backfilling. The drain should be inspected for proper elevation, grade, alignment, and joint spacing; collapsed, broken or cracked pipe; and thickness of aggregate envelope before backfilling. Backfill placement should be in accordance with Subsection 6.10.

7.4.6 Televising. The underdrain should be inspected by means of closed circuit television or other acceptable camera systems where appropriate. Permanent videotape or film should be furnished in accordance with the contract documents.

7.4.7 Testing. There are few definitive tests that can be performed on installed subsurface drains that give measurable indications of the functional effectiveness of the installation.

The specifications may require specific field tests to be performed. As appropriate, samples shall be furnished by the contractor or representative samples will be taken from delivered materials. The number of samples shall be sufficient to satisfy all testing requirements. Control testing shall be performed in the field or at such other locations as required. The inspector, together with the engineer, will review the testing requirements and determine the overall testing program including all needed testing facilities and record forms.

Complete records of the test and results shall be retained. Specimens shall be retained if they are important to prove results of the specified tests. Before being accepted as completed, each drain should be tested for obstructions.

7.5 Acceptance of Construction

Acceptance of construction normally covers the entire job and is not done on an incremental basis. In normal day-to-day operations, field personnel may verbally acknowledge apparent compliance with the contract documents. However, such acknowledgement should not constitute acceptance of part or all of the construction.

7.6 Recording Observations

Work performed by the contractor on a shift basis should be recorded by the inspector to provide a detailed record of the progress. All observations of noncompliance with the contract documents shall be recorded in the daily report. The report should cover any verbal statements made to and by the contractor concerning the noncompliance. Photographs should be taken when they assist in describing the noncompliance.

On matters not immediately corrected, the inspector should give the contractor a separate written Notice of Noncompliance within 24 hours. The notice should state specifically why the work does not meet the requirements of the contract documents.

7.7 Record Drawings

Record details of construction should be incorporated into a final revision of the construction drawings to represent the most reliable record for future use.

During construction, the contractor and/or inspector should measure, reference, and record the

Table 7-1

Inspection Activities Associated with Construction

Construction Activity	Inspection Activity
1. Subgrade/foundation preparation	Observation and verification of grades.
2. Receiving/storing materials	Verification of type and condition of materials received at site and storage procedures.
3. Trench excavation	Observation and verification of width, locations, lines, and grades.
4. Installation of subsurface drainage system	Verification of lines and grades, and testing where necessary.
5. Initial backfilling	Verification of proper filter layer placement.
	Verification of use of proper placement and compaction technique.
	Verification of specified gradations, thicknesses, and densities.[1]
6. Final backfilling	Verification of proper backfill material placement procedure and compaction procedure.
	Test verification of specified density and moisture content.
7. Site restoration	Verification of final site conditions as specified.

---

[1]Samples of filter aggregates for gradation tests should be taken from the layer after compaction.

locations of all inlets, outlets, and stubs for future connections, and other buried facilities. All construction changes from the original plans, rock profiles, and other special classes of excavated material also should be recorded. Contract drawings should be revised to indicate this field information after the project is completed. A notation such as "Revised According to Field Construction Records" or "Record Drawing" should be made on each sheet along with the inspector's name, date, and company name. Records of such plans should become a part of the owner's permanent records.

For subsurface drainage systems, the following minimum information should be included on the record drawing:

- Size and type of all drains on plan and profile sheets.
- Station and pipe invert elevation of all tees, wyes, cleanouts, manholes, and outfalls.
- Manholes and other critical points referenced to established survey points.

Standard Guidelines for
Operation and Maintenance of Urban Subsurface Drainage

Contents

PAGE

FOREWORD

The "Standard Guidelines for Operation and Maintenance of Urban Subsurface Drainage" is an independent document intended to complement the ASCE Urban Subsurface Drainage Manual. This guideline is a companion to the Standard Guidelines for Design of Urban Subsurface Drainage and the Standard Guidelines for Installation of Urban Subsurface Drainage. These standard guidelines were developed by the Urban Subsurface Drainage Standards Committee, which is responsible to the Special Standards Division of the Management Group "F" (MGF) on Codes and Standards.

The material presented in this publication has been prepared in accordance with recognized engineering principles. The standard guidelines should only be used by or under the direction of an engineer competent in the field of subsurface drainage. The publication of the material contained herein is not intended as a representation or warranty on the part of the American Society of Civil Engineers, or of any other person named herein, that this information is suitable for any general or particular use, or promises freedom from infringement of any patent or patents. Anyone making use of this information assumes all liability from such use.

Standard Guidelines for
Operation and Maintenance of Urban Subsurface Drainage

1.0 Scope

The intent of this standard is to present operation and maintenance guidance for urban subsurface drainage. The collection and conveyance of sub-surface drainage waters is within the purview of this standard for applications such as airports, roads, and other transportation systems, and industrial, residential, and recreation areas. Incidental surface water is considered. This document is intended for guidance to the owner during the operational phase.

Some of the applications this standard does not address include agricultural drainage, landfills, recharge systems, storm sewer systems, or injection systems.

2.0 Definitions

2.1 General
This section defines specific terms for use in this standard. The reference documents listed in Section 3.0 may also be helpful in understanding the terms of this standard.

2.2 Terms
Aquifer — geological formation or group of formations through which water flows.

Envelope — permeable material placed around a pipe or drainage product to improve flow and soil filtration conditions in the area immediately adjacent to the drain and for improving bedding and structural backfill conditions.

Geocomposite — geosynthetic materials for collecting and transporting water while maintaining soil stability.

Geomembrane — sheet materials intended to form an impermeable barrier.

Geosynthetic — a generic term for synthetic material or structure used as an integral part of a project, structure, or system. Within this category are subsurface drainage and water control products such as geomembranes, geotextiles, and geocomposites.

Geotextile — a woven or nonwoven thermoplastic sheet material intended to allow the passage of water, but not fines, and without collecting fines at the soiltextile interface.

Inspection — critical examination to determine conformance to applicable quality standards or specifications.

Iron ochre — a red or yellow gelatinous deposit formed by a combination of soluble iron deposits and bacteria.

Jetting — a method used to clean pipes involving high-pressure water.

Observe — to conform one's action or practice, to comply with.

OSHA — Occupational Safety and Health Administration, the federal agency responsible for safety and health concerns on construction job sites.

Rodding — a method used to clean pipes involving mechanical means.

Subsurface water — all water beneath the ground or pavement surface. Sometimes referred to as groundwater.

3.0 References

The references for these Standard Guidelines for Operation and Maintenance of Urban Subsurface Drainage are the same as the references for the Standard Guidelines for the Design of Urban Subsurface Drainage.

4.0 Operation and Maintenence Plan

4.1 General
This guideline is a generalized format of what should be included in an operation and maintenance plan. It may be necessary to modify this plan to more accurately reflect the specific subsurface drainage system under consideration.

This is intended as a general guideline for preparing procedures and timetables related to routine operation and maintenance of subsurface drainage systems and associated components. Technical personnel must be familiar with the basic concepts of subsurface drainage facilities.

Operation and maintenance instruction materials submitted by manufacturers supplying equipment for the drainage system components should be retained and incorporated as needed into the procedures document. The manual should explain the general operational relationships between the various system components of the facilities and include any manufacturers' instructions or recommendations. Adherence to these procedures is essential to retain the operating capacity of a facility throughout its expected service life.

4.2 Scope

An operations and maintenance plan is intended to provide general guidance for the topics listed. However, this list of subjects may not cover all aspects of any particular subsurface drainage system.

- Normal operating procedures to be followed during the majority of the facility's service life.

- Special operating procedures to be followed when normal procedures cannot.

- Routine maintenance procedures specifically described according to a definite schedule.

- Safety procedures for all operating personnel.

- Emergency procedures for conditions involving serious service disruptions.

4.3 Responsibilities

The owner has the responsibility for operation and maintenance of the subsurface drainage system. Generally, these areas of responsibility include:

- Providing adequate funds for supplies and personnel, operations and maintenance of equipment, and any necessary system expansion.

- Selecting either trained personnel or providing training and education for qualified individuals.

- Developing and implementing a comprehensive program that includes a complete reporting and records retention system.

- Coordinating with other departments for maintenance and operation to assure an integrated system is working.

4.4 Design Criteria

The procedures manual should detail the performance criteria used in the design. All personnel should be aware of these criteria so that deviations can be recognized.

4.5 Normal Operating Procedures

Normal operation is the design condition expected of the system. When all components are functioning as designed, the only requirements of the operating personnel are routine checks and scheduled maintenance.

4.6 Abnormal Operating Procedures

4.6.1 Line Blockage. If a pipeline becomes blocked, rodding or jetting may be required to clear it. During clearing operations, a careful watch should be maintained at the downstream manhole for an indication of the cause of the blockage.

Clean-out assemblies may be located at periodic intervals along the pipeline as shown on the "as-built/record drawings." If a conveyance main becomes blocked, the nearest of these clean-outs can be located, and accessed and cleaning equipment inserted.

Hydraulic flushing may be attempted to clean any pipeline. Special care should be taken to avoid damage caused by hydraulic surging.

4.6.2 Other System Components. In the event of reduced performance of system components, the correction procedures set forth in the owner's operation and maintenance manual should comply with manufacturer's recommendations.

4.7 Maintenance Procedures

4.7.1 General. Maintenance can be broadly classified as either corrective or preventive.

4.7.2 Corrective Maintenance. Corrective maintenance involves the repair of equipment after breakdown. The operation and maintenance plan should include all necessary equipment manuals, diagrams, and instructions for satisfactory operation.

4.7.3 Preventive Maintenance. Equipment breakdown is usually related to a failure of preventive maintenance. As the term implies, preventive maintenance is intended to prevent disruptive breakdowns. Since many components that require different preventive maintenance actions at different time intervals may be involved in the system, preventive maintenance is best performed on a scheduled basis from a checklist.

4.7.4 Checklists. Each named component should be checked as required. A current copy of a daily checklist should be put in appropriate locations. Once all inspections outlined on the form are completed, the forms should be replaced and the completed forms filed in the owner's records.

4.7.5 Annual System Inspection. A complete inspection of the entire system should be performed at least annually.

4.8 Safety

4.8.1 General. All personnel are responsible for keeping areas safe and clean. Guards should be in place on operating equipment and all areas should be properly lighted. All enclosed space should be adequately ventilated prior to personnel entering. All personnel should be sure they understand the following:

- Location of all safety equipment.

- Use of safety equipment and devices.

- Knowledge and adherence of safety rules.

- Awareness in "danger" areas.

4.8.2 Mechanical. When working on pumps, suction and discharge valves are fully closed. Maintenance on equipment in operation should be limited to lubrication, packing adjustments, minor repair, or as allowed by the manufacturer's instructions.

4.8.3 Electrical. Lockout and tag main switches of electrical equipment before beginning work. Do not remove any tag without first checking with the person who initiated the tag. Any unusual motor temperature, noise, vibration, etc., should be reported and logged.

5.0 Water Quality

5.1 General

Periodic observation of the flow is necessary to monitor possible water quality degradation. The potential for pollutants to be present is constantly changing. Improvements and/or developments within the drainage basin can generate substances having a pollution potential that could be conveyed to the subsurface drainage system.

An increase in flow rate is often accompanied by higher velocity, which is more effective in transporting pollutants. This, in fact, could be the only cause for concern of such pollutants as suspended solids and/or turbidity. Quantity and quality should be monitored simultaneously.

5.2 Environmental Indicators

A review of the area should be performed to determine any changes since the construction of the subsurface system. These changes will then have to be evaluated as to possible effects on the subsurface flow. Water sampling of aquifers and watershed sources representing existing and potential sources of subsurface water supply may be required. Certain parameters and their background levels can be expected to occur naturally in the water due to the existing environment. By visual inspection or through personal observation, a determination can be made for the necessity and extent of a field sampling program. If test results show unusual concentrations or unexpected constituents in the water, further investigations could be necessary. A treatment program may need to be implemented, or modifications may need to be proposed that would mitigate or eliminate adverse impacts caused by the problem constituents.

5.3 Water Quality Standards

Limitations may be imposed on the discharge based on the receiving waters. Water quality standards limit the concentration of various parameters to be discharged. Standards are generally established by the federal, state, or local governments and subject to continual revisions. These standards can be used as a basis for evaluating water quality. The suggested list of parameters for analysis include, as a minimum, temperature, color, odor, pH, dissolved oxygen, total suspended solids, and turbidity. Parameters for consideration in addition to the preceding list include inorganic chemicals (includes heavy metals), corrosivity, organic chemicals (pesticides/herbicides), microbiological, and radioactivity.

Sampling points and frequency can best be determined in the field. Sampling points should be located at strategic points throughout the system.

6.0 Periodic Inspection

6.1 General

Most systems experience long-term flow capacity reductions. Maximum subsurface drainage system performance requires maintenance programs that keep systems clean, soil-tight, structurally intact, and free of debris.

Routine frequent inspections of all subsurface drainage systems are needed to ensure the continuing level of water management intended by the initial planning and design. A uniform inspection schedule applicable to all sites cannot be established due to regional variations in climate or environment and local conditions affecting land usage. Experiences with similar systems may indicate specific local problem areas that should become a formal part of an inspection program.

6.2 Inspection

Inspections should determine the current operating and structural status of the subsurface drainage system and provide thorough examination for symptoms of developing problems that can alter future performance. Typical elements comprising a thorough inspection are:

1. Check for accumulations of debris, rodents, or other flow impediments at inlets and outlets. Flap gates and other water control devices should operate freely.
2. Inspect the system interior, if possible, for tree or other vegetation roots, mineral deposits, trash or silt accumulations, and other foreign objects obstructing flow paths.
3. Evaluate ground surfaces for evidence of subsurface drainage system leakage. Excessive groundwater from inoperable subsurface drainage systems may cause prematurely distressed pavements, loss of healthy vegetation, and topsoil saturations.
4. Examine inlet and outlet areas for evidence of soil erosion, which generally leads to scour, undermining, and caving of adjacent soils supporting the subsurface drainage system. Soil erosion quickly leads to reduced structural and hydraulic performance.

5. Inspect all visible structures such as catch basins, headwalls, culverts, and outlet pipes, for signs of wear or breakage.
6. Check upstream for evidence of backups or prolonged surface water presence that indicates reduced inflow. Check downstream for evidence of foreign materials that indicate reduced filtration of soils, diminished screening of foreign particles, or structural degradation of the drainage system itself. Visual inspection methods are frequently enhanced by electronic and optical aids such as television cameras, fiber-optic scopes, and laser beam equipment. These inspection tools reveal cracks, displacements and misalignments, and other interior problems with minimal disturbance to the subsurface drainage system.

7.0 Maintenance

7.1 General

Maintenance of subsurface drainage facilities is often neglected. Most of the facilities are out of sight, so there is little public pressure to carry on maintenance programs. Failures tend to develop gradually, as opposed to being catastrophic, so that a failure may not be recognized for several years after it has occurred. Costs of failure in terms of property damage and reconstruction can be enormous compared to the cost of routine preventive-maintenance. Therefore, it is usually in the best interest of the owner to develop and implement an aggressive preventive-maintenance program. Structures such as manholes or outlet pipes should be checked on a regular schedule for signs of structural distress and loss of hydraulic function. Appropriate repairs should be made on a timely basis.

7.2 Cleaning

Methods and schedules for cleaning will depend on the type of subsurface drain being cleaned. The function of the drain, materials from which it is made, susceptibility to clogging, and type of clogging (roots, trash, chemical, or biological) will all need to be considered. High-pressure hydraulic drain cleaners are satisfactory for most applications where access to the drain lines is provided. Initially, annual cleaning should be sufficient for roots and chemical or biological accumulations. As experience with the drain system is gained, the cleaning schedule should be adjusted accordingly.

Some installations, such as sport field drain grids or highway shoulder drains, may not provide access for mechanical cleaning.

Chemical treatment may be required in place of, or in conjunction with, mechanical cleaning. Chemical cleaning must be done in an environmentally responsible manner. For instance, iron ocher deposits, which can clog the openings in the pipe or restrict the carrying capacity of the pipe, sometimes require strong acid solutions for removal. These solutions must be contained until they are neutralized to the approximate pH of the receiving body of water.

7.3 Mechanical and Electrical

Pumps and motors will usually have manufacturer-recommended maintenance guides. When they do not, the design engineer should provide one, particularly if there are any unusual conditions.

Appurtenances, such as covers, valves, or flap gates, may also need occasional attention. Flap gates that are removed for maintenance of the drain should always be immediately replaced. If frequent access is required, a different design of flap gate may be needed to avoid losing the original one.

7.4 Repair

Repairs to subsurface drains should be made using original materials to the extent possible. Since subsurface drains are designed to last up to 100 years, original materials may not be available. When substitutions are made, care should be taken to ensure that the capacity of the drain is not diminished.

The same safety requirements should be in effect for repairs as for original construction. Maintenance crews must avoid the temptation to take short cuts because of the short duration of their activities. Cave-ins and toxic fumes are even more prevalent in repair work than during original construction. Repairs to electrical equipment should be done under standard electrical safety rules.

7.5 Rehabilitation

Rehabilitation of a subsurface drain system may involve complete cleaning of all components; realigning of portions of the system; replacing of worn or outdated electrical components; replacing of gates, valves, and other minor appurtenances; repairing or replacing of major structures, such as manholes and outlets; replacing of entire sections of drains; or additions to the original system.

In general, rehabilitation should follow the same overall pattern as original construction, including engineering analyses, cost comparison, up-to-date materials selection, construction methods, and safety procedures.

INDEX